Marketing Management in Construction
A guide for contractors

Marketing Management in Construction
A guide for contractors

Arthur B. Moore, BSc (Hons), DipEd, FCIOB

Butterworths

London Boston Durban Singapore Sydney Toronto Wellington

All rights reserved. No part of this publication may be reproduced or transmitted in any form or by any means including photocopying and recording without the written permission of the copyright holder, application for which should be addressed to the publishers. Such written permission must also be obtained before any part of this publication is stored in a retrieval system of any nature.

This book is sold subject to the Standard Conditions of Sale of Net Books and may not be resold in the UK below the net price given by the Publishers in their current price list.

First published 1984

© Butterworths 1984

British Library Cataloguing in Publication Data

Moore, Arthur B.
  Marketing management in construction
  1. Construction industry
  I. Title
  624'.068'8   HD9715.A2

  ISBN 0-408-01196-3

Library of Congress Cataloging in Publication Data

Moore, Arthur B.
  Marketing management in construction
  Bibliography: p.
  Includes index.
  1. Construction industry – Marketing – Management.
  I. Title
  HD9715.A2M585   1984   624'.068'8   84-3143
  ISBN 0-408-01196-3

Photoset by Butterworths Litho Preparation Department
Printed and bound in England by Robert Hartnoll Ltd.,
Bodmin, Cornwall

Contents

Preface

Acknowledgements

1 Marketing and profitability 1
2 Marketing for construction contractors 5
3 Marketing strategies 17
4 Marketing procedures and management 31
5 Promoting the company through advertising 44
6 Public relations 62
7 Securing opportunities for contracts 86
8 The contractor as a developer 94
9 Marketing for the small contractor 98
10 Contracting overseas 102

Bibliography 110

Index 111

Preface

Although marketing management is now widely adopted by manufacturing and some service inductries, it has not yet been applied to any great extent by contractors in the construction industry, except those engaged in speculative house-building where its benefits are well understood.

Marketing in its simplest terms means finding out what customers want and then endeavouring to supply those requirements and to do so profitably. The marketing function is an analytical process which precedes selling, although selling is an important part of marketing. Marketing includes market research, analysis, advertising, public relations, company promotion and pricing. It is a very important management discipline which is not clearly understood by those employed in contracting.

This book explains the basic principles of marketing and gives guidance on how they can be applied beneficially by contractors. It draws attention to the interaction between marketing and the other management functions involved in deciding company policy and in achieving operational efficiency.

It is directed to general contractors in the non-housing field and to students seeking qualifications in professional disciplines allied to construction or in management. Much of it will be of interest and use to consultants in construction and to suppliers of materials and equipment to contractors.

ABM

Acknowledgements

This book would have been much more difficult to write without the help and advice of friends and associates. In particular I want to thank my wife who typed the many letters and reports which arose, and Sue Farrow who typed and re-typed the manuscript most efficiently. My thanks go also to Norman Birch and David Seymour of the Department of Transportation and Environmental Planning, the University of Birmingham whose advice after reading the various drafts was most helpful, and to many others whose interest and encouragement helped in no small way.

ABM

Chapter 1

Marketing and profitability

Definition of marketing

Marketing is finding out what customers want and then endeavouring to satisfy the needs and to make a profit in the process.

The Institute of Marketing is more specific and defines it as follows:

> 'The management function which organizes and directs all those business activities involved in assessing and converting purchasing power into effective demand for a specific product or service and in moving the product or service to the final customer so as to achieve the profit target or other objectives set by a company.'

This definition is important for a proper appreciation of marketing.

(1) It is a management function.
(2) It organizes and directs *all* those activities concerned with demand and supply.
(3) It assesses purchasing power.
(4) It converts purchasing power into effective demand.
(5) It delivers to customers.
(6) It achieves the profit target.

What does this definition really mean?

(1) Marketing is a management function, not merely a service to management.
(2) It organizes and directs *all* those activities concerned with demand and supply. What does *all* mean in this context? Does

2

it mean the organization of the company? Does it include staffing and recruitment? Does it mean areas of operation? Does it mean types of work? Does it include pricing?
(3) Does 'assessing purchasing power' mean forecasting demand and, if so, how far ahead?
(4) Does 'converting purchase power into effective demand' mean advising potential customers how best their needs can be met, or how best the company can meet the needs?
(5) Does 'it delivers to customers' mean that it is involved in seeing that contracts are correctly carried out?
(6) Does 'it achieves the profit target' assume there *is* a profit target. If not, does it establish one? How? Does is monitor profit in total or on individual contracts?

All these questions will be dealt with later in this book. For the present, therefore, both a simple and a more involved definition of marketing have been set out, which apply to all business undertakings, wholly or in part.

Before any consideration of the meaning of marketing can be undertaken it is necessary to understand the nature and purpose of a business undertaking. Basically, a business undertaking provides a product or a service to meet a demand or for which a demand can be created, and in the process it hopes and expects to make a profit.

Motor cars are an example of such a provision. There is a demand for motor cars and so firms exist that make them for sale. Because the life of motor cars is limited they need to be replaced, so there is a continuing demand. There are many firms that make motor cars so there is competition, and this leads to the production of new models which vie with each other in the attempts to satisfy the demand. To obtain a larger share of the market manufacturers produce cars with better design, lower petrol consumption and maintenance costs and endeavour to give good value for money. By these and other means they can influence demand for cars in general and their own models in particular.

There are countless examples which could be quoted of businesses providing products, all with competitors providing similar products, and all trying to obtain a larger share of the demand or the market. As demand changes the manufacturer must respond by modifying or changing his products.

In the case of the provision of a service such as banking, insurance or package-holidays, there may be some doubt as to whether the supply of the service or the demand for it came first. It is nearly always possible to create a demand, and many businesses were founded to do just that and have done so successfully. How to create demand will be discussed later.

Business undertakings aim to survive. To do so they must be able to meet their obligations by carryaing out successfully whatever they have contracted to do. They must try to establish a reputation for providing a good product or service at an acceptable price. If they can do this they have the foundation for growth. If they fail to grow they are, in fact, declining because their competitors are growing and leaving them behind. They must so manage their finances that they can use capital for the investment so necessary for expansion. They will also understand that maximizing the quality of life by social investment will not only be good for society but will lead to further increases in profits.

In the public sector the purpose is more complicated but the intention is the same. All public enterprises such as coal, gas, electricity, steel, transport are trying to make a profit. From their profits come investment in new developments and improvements in the products or services they provide and the prospect of greater profits. In some public enterprises there is an element of providing an essential public service at a price the public is prepared to pay, so making a profit is much more difficult and subsidies from public funds are necessary to enable the service to be maintained. There is, however, no doubt that these enterprises are trying to make a profit, but in some cases, e.g. transport, this is extremely difficult without an unacceptable reduction in the services provided.

In these enlightened days, there are some people who believe that profit is not the prime aim of a business undertaking. They go so far as to say that the creation of employment and the provision of career opportunities for employees are the prime objectives. Judging by the extent that overmanning exists or existed in some industries it would appear that the profit motive may have not been given its proper prominence, and that those in control may subscribe to these alternative views of the prime purpose. How wrong they have been is now evident. If a business succeeds in making a profit it will expand and create additional employment and as employment grows so career opportunities for its employees will develop.

It is now widely recognized that marketing planning is an essential part of management and most large and medium-sized companies have marketing directors and/or marketing managers. Research into market demands has a very pronounced influence on the nature of products and services and these have to be tailored to meet customers' requirements.

It is important that it is clearly understood that the purpose of marketing planning is to increase profitability. Careful consideration of the available and potential markets and those that can be created is the basis on which future profitability is built. Unless the opportunities and the ways in which they are likely to change are clearly recognized the vital decisions necessary to ensure growth and increased profitability are unlikely to be made. This is a function of managerial attitudes which must start at board level and be followed throughout the company. Too often opportunities are not recognized soon enough for the full benefits to be obtained, and sometimes competitors have already made their moves.

This brief explanation of the meaning of marketing should alert contractors to the need to consider how marketing planning can help them to manage their companies more successfully. Detailed guidance on how they might do so is covered in the chapters which follow.

Chapter 2

Marketing for construction contractors

It is interesting and informative to note that a Working Party on Building appointed by the Minister of Works submitted a comprehensive report in 1950 which covered almost every aspect of the industry. Nowhere in that report does the work 'marketing' appear. Indeed, it was not until the late 1960s that there was recognition by some sections of the industry, particularly by the National Federation of Building Trades Employees, that the construction industry might benefit from a study in depth of marketing techniques. There was, and still is, some confusion in the industry as to what marketing means, and many have formed the opinion that the techniques cannot be applied to the construction industry.

In July 1970 a working party of the National Federation of Building Trades Employers published a report on 'Marketing for the Building Industry'. Although it did not come out with any clear recommendations as such, the report raised a number of questions, and important questions too, which individual firms needed to consider and answer in respect of their own firms. This was a good report and ought to have stimulated thought in those firms which studied it. It gave some prominence to market-orientated planning based on consumer needs and not on a succession of jobs approached on an *ad hoc* basis. This recognition that the needs of the market should be the basis of a firm's policies and not its current construction ability is most important.

Contractors in the immediate post-war years were faced with demands which they could not meet with anything like the speed or efficiency which clients would have liked. This was understandable enough. The industry had declined in manpower from 1 362 000 in 1939 to about 500 000 in 1945 and then increased quickly to about a million eighteen months later. Much of this additional labour was not fully trained or experienced and management efficiency was affected by these deficiencies. These problems were not confined to the contracting side of the industry. The architectural, engineering and quantity surveying professions had similar problems of lack of training and experience. Furthermore, there were inadequate supplies of construction materials and components as industry began the transition to a peace-time market.

The vast demand for constructional services made the continuation of war-time controls inevitable. These were difficult to operate but achieved a large measure of success until it became clear in 1954 that they were no longer necessary. With their abolition came ambitious redevelopment schemes for industry and commerce and plans for public authorities, often made years earlier, were translated into actual construction contracts.

Since 1954 there was a continuous demand for construction services for new buildings of all kinds and for motorways, roads, and infrastructure. Contractors found little difficulty in operating at desirable levels, although inevitably there were fluctuations in demand. Apart from planning permission and the requirements about distribution of industry there was no direct control, nor would an attempt to exercise such control have been welcomed or acceptable. The result was that, occasionally, the industry was overloaded for a fairly short time or it was underloaded. These almost inevitable variations in demand resulted in much comment but not much action. It is fair to say that the industry increased its output in real terms not so much by a straight-line growth as by a saw-edged upward movement with variations in demand, upwards or downwards, from time to time.

In the early 1970s, in an effort to halt rising unemployment and economic stagnation, the government significantly increased public expenditure on construction, lifted the ban on office building and began to apply industrial development controls less rigorously. These policies resulted in a level of demand considerably in excess of the industry's capacity. This overheating lasted until 1973 and its effects took a little longer to subside, so that it was not until 1974 that contractors started to become seriously concerned about the level of demand.

In the 1980s there has been a distinct and substantial reduction in demand resulting from reductions in public spending as a policy to try to combat inflation and to transfer resources from the public to the private sector. Demand in the private sector also decreased because of high interest rates and there was a drop in home demand because of rising unemployment. The net efect of these two reductions in demand was a serious lack of orders for the construction industry, very severe competition for the work available, and a high level of unemployment. Prices decreased substantially and consequently profits also fell. It is not clearly understood that the bulk of the orders resulting from public expenditure are carried out by firms in the private sector, and that public investment cuts cause reduced investment by private sector companies as demand for their products or services declines.

These circumstances have led to investigations as to how turnover and profitability in the construction industry can be improved and marketing has been prominent in these considerations.

To sum up, since 1945 until the last few years most contractors have not had insuperable difficulties in obtaining an acceptable turnover and profit level in spite of the inconsistency of demand which has varied in a cyclical way from excessive to insufficient. These variations have been almost inevitable in a complex economy, affected as it is by external as well as internal events. Bankruptcies have been high for a variety of reasons – insufficient capital, bad management, slow payments for work done and bad luck – but most firms have survived. Continued survival depends on securing an adequate part of the available market and this means good management – and good management requires market planning.

This is not to imply that marketing management is only of value when the total market demand is low, although low demand obviously prompts firms to investigate how they can contrive to obtain a larger share of what is available. Marketing management is just as crucial when there is excessive demand when contractors are, as it were, spoilt for choice. But they have to choose and it is important to choose in the best interests of the firm for the future, and that means for at least a few years ahead. Those firms that have for some time adopted marketing planning will be better placed to deal with falling demand than those that have not. A recession will cause some firms to adopt marketing planning for the first time; the wiser firms adopted it much earlier.

Although the basic principles of marketing are becoming better known, many still believe that marketing is selling and nothing else, and even senior directors of old established and successful companies hold on to this view. Why is the construction industry lagging behind other industries – both consumer goods and service industries – in accepting the benefits of marketing? Amongst the reasons are:

(1) Contractors have not previously met insuperable difficulties in obtaining adequate levels of work to ensure survival.
(2) The innate conservatism of the industry.
(3) Contractors believe the most important part of their organization is the production side.
(4) The industry is not capable of being planned.
(5) Contractors do not really understand the meaning of marketing.
(6) They do not appreciate that its adoption can lead to increased profitability.
(7) They consider that only clients create demand and that contractors cannot do so.
(8) They believe that good and reliable service will ensure a satisfactory level of orders.

These reasons are put forward in defence of methods which contractors claim have been the basis of their successes in the past. Most of them are invalid for one reason or another:

(1) Demand has decreased enormously in the last few years and seems unlikely again to increase to the levels of the early 1970s. Contractors are not finding contracts easy to obtain and consultants are receiving requests from up to 100 firms to be included in tender lists, when all they need is six or eight.
(2) Contractors are by nature conservative. They accept change only slowly and so new methods of management and new outlooks exist a long time before they are accepted. This may be a result of the nature of their work. From the birth of an idea to carry out development to its completion requires a long time because of the hurdles to be jumped and the detailed planning to be undertaken. New methods also take a long time before they are adopted in the sure knowledge that they will lead to improvements. Examples of too hasty decisions taken with insufficient evidence to support them are not difficult to find. Industrialized building schemes, high-rise flats, curtain walling, excessive use of glass, electric underfloor heating are examples which come readily to mind where the results have been disappointing to say the least.

(3) Many firms are production-orientated, and it is not difficult to understand why. Most, if not all, construction firms started in a small way undertaking small contracts with a handful of staff, a little plant and transport, and a few operatives. As they grew, the principals or proprietors, who may originally have been craftsmen, remained in control and their skills were in carying out the work. Although they engaged others to help as the firms grew, the founders remained in charge and there was no obvious need for any serious consideration of management changes likely to be beneficial. In short, the firms were run by men with little or no training outside production and so the views of the production side had an overwhelming influence on policy decisions.

(4) In an industry so accustomed to dealing with consultants' plans and to making their own production plans using, in some cases, critical path techniques it is difficult to understand why they do not accept the need for planning the whole operations of the individual companies. The argument that overall demand can and does change quickly as a result of events entirely outside the industry's control is not confined to the construction industry. There are factors affecting demand for all industries and the changes are not all in one direction; there are ups and downs and always will be. This reinforces the need for marketing and planning.

There is a feeling amongst contractors and indeed within the Federations and professional institutions, that the industry is used by government as an economic regulator to suffer reduced demand when times are bad and to be overloaded when times are good.

It is said that it is easy to reduce demand by simply reducing the funds available for construction. Demand can be reduced by simply instructing the Property Services Agency of the Department of Environment not to place any more contracts, or to slow down the placing of contracts for certain types of government projects. There is something in this, but it is by no means the whole story. The chances are that government departments are asked to make reductions in their overall spending (which covers other things besides construction projects) and do so to the best of their ability. It may well be that the reviews undertaken by departments eventually mean that some proposed new construction projects will have to be postponed or cancelled, but that would be a consequence of some other reduced or changed activity for which the department was responsible and which would affect other things besides construction. Since almost every activity one can think of involves the construction industry at some time it seems clear that whatever decisions to reduce public expenditure are taken will reduce demand for construction services. However the construction industry will not be the only industry to suffer a reduction in demand; so will the steel industry, the materials industries, the power industries, the telecommunication industry, the furniture industry and a whole lot more. It is not correct to believe that government is attacking the construction industry; the contraction in the industry is a consequence of reductions in other activities. In fairness, it is a fact that when the economy is flagging, the construction industry is under-employed, but so are many other industries. Equally when the economy is buoyant the construction industry is busy and so too are many other industries. Indeed, the state of trade in the construction industry is a good indicator of the state of the economy.

This misconception about changes in government policies and how the effect is transferred to the construction industry should be clarified so that contractors will take a much more careful look at likely government policies than perhaps they have done. Notice that the expression used is, 'How the effect is transferred', and not 'directed at', the construction industry.

The speed at which the effects are felt is an important difference between decreases and increases in demand. Decreases, especially in the public sector, can and sometimes do occur very quickly. All that is required is an instruction to place no more contracts in particular sectors and it is not unknown for work on projects under construction to be stopped, although this leads to increased costs through compensation to contractors and when the project is permitted to continue at a later date. In both the public and private sectors it can mean that schemes in various stages of planning are shelved or cancelled. These reductions in demand have a serious effect and if continued for an appreciable time cause severe problems for contractors who depend so much on cash flow.

When the situation calls for increases in demand, these cannot be translated into contracts with anything like the speed at which demand can be reduced. Even work stopped whilst under construction cannot always be re-started months later without some delays. So far as projects shelved in the

planning stages are concerned, they cannot just be taken forward as if no delay had occurred. There is nothing as effective as the passage of time to inspire second thoughts, and this frequently means the revision of the previous drawings, bills of quantities and costs, all of which can slow down the progress towards contracts.
(5) The meaning of marketing is gradually becoming clearer as a result of the increased number of courses and discussions on the subject, but more understanding of the techniques and therefore more teaching is still required.
(6) Contractors need convincing that marketing planning produces beneficial results in terms of profitability and growth. This will be best demonstrated by the failures of those firms which do not accept it and the successes of those who do, and word spreads quickly in the construction industry.
(7) There is, today, ample evidence that contractors can create work for themselves, even outside the speculative housing field. Contractors can become developers and many have, especially in the industrial and commercial building fields. This is dealt with in chapter 8.
(8) Good and reliable service will certainly lead to some orders, but old and loyal clients cannot any longer be relied upon not to seek competition, something they may have neglected to do in the past. It is a very competitive market these days and opportunities for orders have to be pursued vigorously in competition with other firms who also give good and reliable service.

Since 1970 marketing has been widely adopted in most industries, both manufacturing and service. The Industrial Training Act 1964 encouraged the provision of training courses of all kinds and many of these were about marketing. Firms needed little encouragement to send staff on these courses because they received grants from the training levies which all but paid the course fees and expenses of those attending, and marketing courses were frequently over-subscribed. Strangely enough, or perhaps not so strangely, courses on marketing specifically for the construction industry were not available, partly because there was little demand for them. It is only in the last two or three years that contractors have recognized that they must turn their attention to marketing and now marketing conferences, seminars, and teach-ins are becoming more prevalent. At last the industry is beginning to view marketing as an important management function.

12

It is important to appreciate that marketing in a service industry differs considerably from marketing in a consumer goods industry, although the basic principles are the same. In each case the objective is to increase the profit level, i.e. the return on capital, and basically this is most likely to result from increased turnover. Increased turnover leads, or should lead to a reduction in prices since overhead costs are spread over larger outputs, and price reductions achieved in this way should lead to even greater turnover. The speed at which growth can be attained depends on many factors, not the least of which is the availability of finance. In this connection the construction industry has always been undercapitalized, and suffers greatly from delays in payments for work satisfactorily completed.

Consumer goods industries have long undertaken market research as a means of assessing demand and of the desirability of change. These surveys, undertaken on a sampling basis by the companies' own marketing staff or by specialist market research organizations, have helped in establishing what customers want, what they like or dislike about products and so on. The construction industry would find it very difficult to indulge in this sort of activity since it does not itself often manufacture products, but builds whatever clients and their consultants require. Nevertheless, some firms in the speculative house-building field have resorted to market research to help them decide what types of houses, size of rooms, fitments, etc. prospective purchasers would like, and have incorporated the results in their designs. Furthermore, consumer goods industries can advertise specific products at specific prices and do so with some success in the press and on television. Often the results of these promotional efforts can be seen in almost immediate increases in demand for these products. Construction companies can similarly advertise the services they provide, but it is certain that there will be no immediate flood of enquiries for those services from any prospective clients.

Consumer goods industries are in the fortunate position of being able to compare their prices with those of their competitors selling similar products and can make such adjustments as they consider necessary. Contractors can only compare prices with those of their competitors after the event, i.e. after tenders have been submitted, and then only if the client makes all the prices available, a practice which is less frequent than it used to be in the public sector and almost non-existent in the private sector. Even when competitors' prices become known the information is not as valuable as it might be because it is not possible to ascertain where the variations occur. To an unsuccessful tenderer the disclosure of prices simply indicates the differences between the various tenders. It does not indicate how much has been included for materials, plant and labour or overheads or the amount included for profit. So the availability of competitors' prices is not of great help to a contractor, except, perhaps, as an indicator of the intensity of competition.

It follows that consumer goods industries are in a very good position as a result of market research to make a product, the demand for which can be quite closely estimated and priced so that it will sell. What is more, its design can be altered to meet customers' needs, or to make it more competitive with similar products of other manufacturers. The contractor outside the housing field can do none of these things. He has to build according to the plans and specification of consultants, and he has to price their requirements. There is little scope for market research in this context.

The traditional method of placing contracts is to present plans and specifications and a bill of quantities to contractors and to ask them to submit prices for carrying out the work, usually within a stated construction period. In the case of open tendering, it could be argued that the sole criterion on which the successful contractor was selected was the lowest price. Whether this was ever the ideal way of placing contracts is extremely doubtful because there was no way in which the capabilities, the resources, the experience, the financial status or the quality of work of the successful contractor could be ascertained beforehand. In circumstances of this kind, how can marketing help? The answer is, precious little, except to ensure that sufficient forthcoming contracts are known about to enable an adequate number of tenders to be submitted, and that the proportion of successes will produce an acceptable turnover.

Fortunately, open tendering, except for very small projects, is now a thing of the past and approved lists of contractors are kept by most public authorities and by a good number of private clients and their consultants. To be included in an approved list a contractor has to complete a detailed questionnaire relating to the company. The questions asked vary from organization to organization but are very searching, and even nationally known firms of high reputation are not exempted from this treatment.

Before a company is given the opportunity to submit a price it has been thoroughly vetted by the client or his consultants as have all the other firms included in the tender list. It follows that price is no longer the sole criterion, because a firm does not get the opportunity to quote unless it has met the criteria of suitability, competence, experience, resources and stability which determine whether it will be considered for inclusion in the tender list. So there is an opportunity to promote a firm by the way the questionnaires are completed and particularly by the contents of any covering letter which accompanies the return of the completed form.

But competitive tendering as described above is not the only method of placing contracts. The Banwell Committee Report stressed the alternative ways open and advised that each project should be dealt with in the way that seemed most appropriate. It is the availability of these alternatives which widens the scope of marketing activities for contractors.

The structure of the construction industry

Some of the reasons for the reluctance of contractors to accept marketing planning have been examined, but perhaps the contracting industry is fundamentally different from all other industries. To see whether this is so, it is necessary to examine the structure of the construction industry so far as the contracting side is concerned.

Construction work is mainly carried out on sites, most frequently in the open air and in varying locations. Although there have been extensive developments in the manufacture of components away from the site, and this practice will no doubt continue to increase, the principal work for the contractor will be performed on site in excavating, building and assembling materials and components. To this exent it is very different from most industries where work is carried out under cover at a permanent location.

In general, construction work falls into two categories, building and civil engineering, but projects frequently are a mixture of both. Foundations, drainage and roads occur on most building sites and can be classified as civil engineering whilst structures above ground and, in some cases below ground, can be classified as building. Building can be subdivided into four main types:

(1) *General contracting* includes public works, industrial and commercial building where it is usual for the client through his consultants to produce the design and specification and for the contractor to provide the construction services, using sub-contractors where necessary.

(2) *House building* is carried out in both the public and private sectors. In the private sector the contractor is frequently involved in the design as well as the erection function and in selling the product. Marketing in the private housing sector is accepted as a matter of course, and in consequence is not considered in this book. Indeed, a complete study of marketing in private house-building warrants separate treatment.

(3) *Maintenance, repairs and modernization* is growing in importance, particularly modernization or refurbishment. Many old buildings can be adapted to modern standards far more cheaply than the cost of constructing new buildings. What is more, this work can frequently be carried out with little or no interference to normal activities within the building. Maintenance and repair contracts may be either single projects or annual contracts in which individual jobs are of relatively small value but the total value of the annual contract is substantial. In these cases the need for prompt service is

Table 2.1 Numbers of firms October 1981

General builders	45 889
Building, civil engineering contractors	2 880
Civil engineers	2 121
Plumbers	9 866
Carpenters and joiners	6 997
Painters	12 703
Roofers	3 483
Plasterers	2 975
Glaziers	2 164
Demolition contractors	458
Scaffolding specialists	512
Reinforced concrete specialists	367
Heating and ventilating engineers	5 634
Electrical contractors	9 187
Asphalt and tar sprayers	630
Plant hirers	3 508
Flooring contractors	924
Constructional engineers	1 092
Insulating specialists	847
Suspended ceiling specialists	502
Floor and wall tiling specialists	709
Miscellaneous	1 738
Total	115 186

Source: *Housing and Construction Statistics 1971–1981*, HMSO

Table 2.2 Number of private contractors by size group October 1981

Size group	Number of firms
1	40 580
2–3	34 541
4–7	20 187
8–13	9 161
14–24	5 380
25–34	1 791
35–59	1 721
60–79	528
80–114	416
115–299	598
300–599	162
600–1199	82
1200 and over	39
	115 186

Source: *Housing and Construction Statistics 1971–1981*, HMSO

It is clear from these figures that more than 90% of firms employ fewer than 14 operatives, and only 121 employ 600 or more. So the industry is mainly comprised of small firms. This may be a reason for the neglect of marketing planning. It is hoped to show in this book that it is not a valid reason. The application of marketing principles is independent of the size of firm but the precise methods of application vary for firms of different sizes.

such that labour must be immediately available to deal with emergencies. For this reason these contracts are not particularly popular with medium or large companies, but they should not be overlooked. Annual maintenance contracts for industrial companies for work to be carried out in shut-down holiday periods are lucrative and well worth pursuing.

(4) *Sub-contracting* is much more prevalent now than it was before the Second World War, partly because services required are much more sophisticated. The extent to which sub-contractors, other than those nominated by consultants, should be employed is a matter which should be considered along with other matters which affect prices and quality.

The various methods of obtaining orders are dealt with in detail in chapter 7.

It is important and interesting to examine in some detail the structure of the contracting industry. *Tables 2.1* and *2.2* show the numbers and types of private contractors in October 1981 and the numbers of them which employed various sizes of workforce.

Chapter 3

Marketing strategies

For existing firms of all sizes it will, without doubt, be the policy of the controlling authority – principals or board of directors – to survive and to grow, and one of the first things which must be decided is what sort of growth is both desirable and possible, and at what rate.

What is meant by growth? Is it more jobs, is it more employees, is it more turnover or is it more profit? These four possible measures are interdependent. If there are more jobs, i.e. orders, it is likely that more employees will be needed on the staff and on site. If there are more projects, turnover is likely to increase. It does not follow that increased turnover automatically means more profit, but it will do if a good standard of efficiency can be continued or improved. There is a danger here. In order to obtain greater turnover, there may be a tendency to lower prices so as to secure more contracts. The profit increase would then result from lower returns on a greater turnover. This needs very careful consideration because a few contracts which turn out to be unprofitable could upset all the calculations, and although turnover would be increased, profit might well decrease. There have been examples of this sort of policy in recent years. When turnover decreases because of a high level of competition (which always arises when total demand or demand for a particular type of work decrease), there is sometimes a determined effort made to increase turnover by taking contracts at unrealistic prices. Although turnover and cash flow increase and employment levels are maintained, profit decreases. A good example of increasing or merely maintaining turnover at the expense of profit is the scramble for new motorway work. There are some 20 or 30 contractors experienced on motorway construction, all of them with large holdings of the types of plant and transport required for these types of contracts. The reduction in the availability of motorway work in recent years because of cuts in government spending and delays because of protracted public enquiries to deal with objections, resulted in a fierce determination to win whatever contracts were available. The inevitable result was that prices were reduced well below realistic levels and some contracts were started at prices much below the Department of Transport's own estimates. Whether any profit could possibly be made by the successful contractor was extremely doubtful.

In considering growth and what it means, contractors should be very careful not to regard it simply as more turnover, more employees, bigger premises. There is only one safe way to plan growth and that is to plan for increased profit on capital employed. If that plan succeeds, it is more than likely that the other criteria will follow.

The relationships between management and employees have undergone quite revolutionary changes in comparatively recent times. Because of legislation introduced to bring a greater sense of fairness into industrial relations, employers are now much more thoughtful about reducing manpower or taking on additional employees than they were 20 years ago. Industrial relations in this country are, in fact, very much better than one is led to believe by the media. The UK is way ahead of many European countries and has many fewer rules and regulations, but paying off employees can be expensive, and taking on new employees has a sizeable effect on staff costs. So managements tend to be cautious in deciding to increase or decrease their employment levels. There is, therefore, a tendency not to dismiss employees unless it is absolutely unavoidable.

The danger inherent in this reluctance is that a policy will be followed of trying to obtain sufficient work to keep current employees in work. Laudable as this policy is, it is not sound for two main reasons. First, it means that employees are kept on where there is insufficient work to justify their retention. This has a cumulative effect. Their retention increases overheads or leads to uneconomic manning of sites, or to the carrying out of unnecessary tasks created to give employees something to do. On the one hand it tends to increase tender prices and so reduce the prospects of obtaining orders, and on the other it makes it less likely that a satisfactory profit will be made on work in progress. Large firms can follow this policy for a time, but the risks are that unless the availability of work increases and competition becomes less keen, the work load is likely to decline still further and with it profitability as prices are lowered in an effort to secure more contracts. In such circumstances the number of employees for whom no worthwhile or essential work exists will increase and the problem becomes even more difficult to resolve.

Secondly, the question of unused plant and vehicles needs careful consideration. Idle plant and vehicles are to be avoided as far as possible and consideration must be given to the advisability of retaining their operators if there is no work for them.

The foregoing paragraphs are not suggesting, as some might think, a ruthlessness in dealing with employees which no contractor with any humanity would adopt. They are statements of some of the considerations which must be made before any sound judgement of the way forward can be determined. What is more, they point out clearly that growth is best measured by profit and that the success or otherwise of a firm is judged by its profit on capital employed – not on turnover, not on projects well built, not on number of employees, not on any other criterion. These other criteria help to build up the image of a firm, but profit is what they must place firmly at the head of the list.

Accepting that profit is the prime objective of any firm, (after all, that is why the firm came into being in the first place), there are other considerations today which will help to achieve the main one, and which are essential for the survival and progress of the firm.

1. Satisfaction of shareholders' interests

Most firms rely on investment by the general public or in the case of small firms on bank support and any loss of confidence on their part can have serious consequences for the firm.

2. Retaining the confidence of suppliers and sub-contractors

Promptness in paying creditors is as important to contractors as is the speed with which their creditors pay them. Although credit controllers exist in most firms these days, they are concerned with money owed to the firm. It is just as important to ensure that their own debts are discharged promptly, and firms that do so will benefit from the confidence generated in their suppliers and sub-contractors, and the improved service that will follow.

3. Satisfaction of employees

In these enlightened days employees are no longer just numbers. They are individuals of intelligence, often highly qualified for their work and have a right to be treated as such. They will not expect favours, or most of them will not, but they will expect a sort of rough justice. They must be kept as happy as possible and given proper assistance with their problems at work and with any personal problems on which they may seek advice. Staff associations are well worth considering. Meetings of management and staff representatives should be held from time to time to inform and discuss matters of interest to the staff who should be encouraged to bring forward for discussion any items which may be causing them concern. If there is a personnel manager he is best to chair these meetings, and the vice chairman should be chosen from the staff by the staff themselves.

4. New fields of endeavour

There are many ways in which expansion can be attained and these will be discussed in some detail later. It is sufficient to say here that new areas of operation or new types of work to be pursued need the most careful consideration before decisions are taken. The policy change should come after this consideration and not before it.

Contractors will need to take stock of the current state of the firm. What is the turnover and the profit? Has it come about as a result of acceptable profit levels on most projects or was it achieved because of very good profits on some jobs and poor profits or even losses on others? Were there any types of work where profits were considerably better than on some other types?

These and many other matters need to be looked into carefully before a decision is taken about possible future budgets or profit targets. To enable these examinations to be undertaken, there must be adequate and easily readable job records, a field where a good deal of thought is required to ensure that job records contain all the information needed and that the information has been properly recorded as soon after the event as possible. It is surprising how often impressions of what actually happened are shown to be unreliable by reference to well-kept record cards, and a model record card will be discussed later.

21

Factors that affect demand and supply

Demand for construction work, like demand for products and services of all kinds, is a variable which changes for many reasons.

(a) Political decisions and legislation

Public authorities account for roughly 40% of total demand for new non-housing work. It follows that sizeable variations in total demand can result from variations in demand from public authorities. The client with the biggest annual demand is central government, and this demand is translated specifically into work for the construction industry or direct employees of public authorities by the various government departments. The Property Services Agency and the Department of Transport are the major clients, but there are other departments who place work through agencies. Changes in government policy as a result of either internal or external matters or events can and do affect the demand for construction services from those departments.

(b) Industrial and social factors

The decision of OPEC to increase the price of oil had a far-reaching effect on energy policy. It increased costs of energy for transportation, lighting, power, heating and the cost of oil-based products. Since that shattering event, energy needs and methods of meeting them have never been out of the news for long. The whole question of oil supplies for the future has been examined and re-examined and the offshore oil supplies have made a significant contribution to the problems so far as the UK and some other countries are concerned. Indeed, offshore oil exploration and drilling in the UK created a demand for construction services of all kinds: Aberdeen became a hive of industry with huge increases in demand for workshops, housing and all the attendant requirements of people, and Aberdeen was by no means the only part of the UK which gained in employment opportunities. For example, specific equipment has been needed for the offshore exploration and drilling. This has created demand in engineering and other industries, more than half of which has been met by British companies.

The oil price increase also had the effect of making it essential to look to alternative sources of energy. The coal industry has put forward plans for new mines to produce good quality coal, with sufficient reserves to last for many years. These plans, when approved, will provide work for the construction industry for years to come.

The generation of electricity by the use of nuclear power has been under consideration for many years and decisions have now been taken to build more nuclear power stations. The use of water power by schemes such as Dinorwig in North Wales to provide electricity, and the possible use of tidal power are also examples of considerations which could lead to demand for construction services. Yet a further example is the possible use of giant windmills along parts of the coast to generate electricity.

Another factor affecting demand for the services of the construction industry has been the rapid growth in the electronics industry, an area still in its infancy. The development of computers has revolutionized many of our industries and commercial activities. What formerly was time and labour consuming can now be carried out extremely quickly by electronic means and with relatively little labour. This has created demand for manufacturing facilities and for computer centres on a very big scale, and this demand will continue. As the electronics industry progresses, new generations of equipment will be required to replace older installations and this will be a continuing process. Automation often controlled by electronic means has revolutionized substantial parts of manufacturing industry.

Whatever the real causes of our economic troubles, the growth in unemployment is having its effect on construction demand. The introduction of the five-day week increased demand for leisure facilities, and unemployment, naturally and unfortunately, leaves more people with time to spare. It seems more likely that more leisure facilities will be needed and this will result in demand for sports centres, swimming facilities and, no doubt, squash courts where availability at present nowhere near meets demand.

The social changes in the UK have been very considerable indeed. The growth in the package holiday business has resulted in demand for more flights, more sea crossings and this means demand for runways, hard standing, airport buildings and for better dock and landing facilities at seaports.

Safety measures following legislation for higher standards of safety at work and at leisure have created yet further demands on the construction industry. Whether they be fire regulations in buildings, safety regulations in factories or at football grounds, work has had to be done in numerous places.

The development of containers for moving goods around the UK and overseas and the increase in size and weight of goods vehicles have also created demands for the industry. Storage and hardstanding, roads and by-passes and port facilities have all needed development or improvement and in the case of motorways, extensive rebuilding and repairs.

The fall in birthrate a few years ago has meant that the school population has decreased and so a review of the number of school places has been necessary, along with a review of the level of teacher training facilities needed for the future. What effect this will have on the construction industry is not at present clear, but demand for new schools and colleges could decline somewhat.

In contrast, where new housing estates have been developed, the need for schools, shops, places of entertainment and other facilities has become evident.

The much greater use of the motor car for shopping and leisure has changed ideas about where shopping facilities should be provided. There has been a marked increase in the tendency to move away from town centres for shopping, even on Saturdays, and facilities in the suburbs have increased. This is remarkably so in the case of supermarkets and department stores and the service industries also have opened offices and showrooms in the suburbs. This trend has increased the demand for car parking facilities and surface and multistorey car parks have been provided to meet this demand, but more are needed.

One of the most remarkable changes in habits is the increase in dining out as a means of entertainment. The greatly increased food trade of public houses and hotels is evidence of this trend. Quite enormous car parks have been created to enable those who require meals to park without trouble. In many cases, to provide the necessary dining and cooking facilities, large extensions to the public houses and hotels have been undertaken; this has also been the case with clubs and other establishments where refreshment and entertainment go hand in hand.

Package holidays have not been confined to overseas, although this is where the growth has been greatest. In the UK, hotels offer attractive packages such as 'bargain breaks' for week-ends or longer periods. This has not necessarily meant extensions or alterations, although these have been undertaken in many hotels. The open-air pools at nearly all the hotels in the popular continental resorts have induced UK hotels to follow suit, and many seaside hotels now offer this amenity where bathing in the sea was formerly the custom.

These examples of changes in social habits are not comprehensive, but they give an indication of trends and how, with a little thought, growth in demand for construction services can be foreseen.

A government publication entitled *Social Trends*, published by HMSO, is a good source of information about changes in social habits and is well worth studying.

(c) Ecological pressures

One of the advantages of a democracy where free speech is permitted and encouraged is that the views of minorities can be brought to notice and dealt with in a civilized way. When planning permission is required before any sizeable construction project may proceed, there is every opportunity for objectors to state their views and to have them carefully considered before final approval or refusal is given. Proposals for new motorways and roads, new airports, nuclear energy stations, waste disposal and the possible pollution of rivers, new coal fields and many other types of project have all been subject to lengthy enquiries before approval was given or refused. Few in number but vociferous objectors have frequently managed to ensure that proposals were amended or even cancelled and contractors might well benefit by keeping an eye on proposals that are likely to result in demand for their services, but which are also likely to be contested at planning stage. Certainly marketing directors must watch the various stages closely and, as soon as it is clear that a go-ahead is likely, they should pursue the project with the client or his consultants, preferably after having expressed an interest in the project at an earlier stage.

(d) The international scene

The world is a small place today with fast transport services available. Communications by radio and television are also fast and events far away are known almost at once all over the world. It would be very nice if it were possible to concern oneself solely with the events in the UK. Unfortunately, this is not the case. Quite apart from our membership of the European Economic Community and the North Atlantic Treaty Organization, we are quite deeply involved in what is going on in most other parts of the world. What goes on in other countries usually has some effect on what happens here and as a relatively small country which relies on overseas trade for its standard of living, it would be very surprising if this were not so.

This book is not intended to be a primer on economics, but it does seek to remind the reader of those economic factors that will affect demand in the construction industry, such as the way in which the domestic policy of the United States has a great effect on the economy of the UK. High interest rates in the US mean high interest rates here, and these in turn reduce UK demand and are a disincentive to entrepreneurs so far as development and expansion are concerned, and this affects the construction industry.

Defence considerations in conjunction with our allies or without them affect demand for construction, not only for service accommodation and facilities, but also for those industries that provide the equipment needed by the armed services.

Tariffs and trade restrictions and subsidies affect demand for British products and this in turn affects demand for construction.

These are only a few examples of the factors external to the UK that have repercussions in this country and which ultimately affect demand for construction services. It follows that marketing directors must keep in touch with policies not only of the British government, but also of overseas government if they are to make reasonably accurate forecasts of demand levels for construction services.

(e) Inflation

Everyone talks about inflation as though it is something new. The truth is that it has been with us for a very long time and probably will be for ever. With the exception of a few years between the wars when purchasing power was low, the value of the pound has been declining. What has been worrying is the rate of decline which has been very high but has now decreased.

Contractors should always look at the real value of turnover, profit and orders in order to make valid comparisons with past performance and planned expansion. (To obtain real prices deflate current prices to what they would have been in a base year by using the *Index of Building Costs* published by the Department of Environment.)

The foregoing gives some examples of influences on the future demand for constructional services. They show the sorts of things which can increase demand in particular fields and those which can reduce demand. The list is by no means comprehensive and marketing directors will need to read extensively and make their own judgements.

In large companies or where a number of companies operate as a group a corporate plan will be needed. This plan will have to take account of the plan for the general contracting part of the organization. These two plans are interdependent, and what follows is an indication of action necessary by the general contracting company, whether it is part of a group or a single company.

Marketing planning is a continuous process. The marketing director or manager is constantly reviewing likely demand through his reading and his discussions with people from many walks of life whose views he must seek and consider. He is, as it were, the intelligence officer of the firm making as sure as he can that he is well informed about policies, trends and proposals which could affect clients' demands. He is concerned about the nature of demands, the location of them and the timing. He will find this is a more than full-time job, but should be able to arrive at conclusions accurate enough for planning purposes.

Once the decision has been taken to accept the philosophy and techniques of planning the way forward, the next question concerns the methods to be used. Clearly these will vary according to the size of the firm, its location and the location of its outstations if any, its resources and the types of contracts it undertakes or could undertake. For small firms, the problems to be solved and the decisions to be taken are unlikely to be very difficult. Decisions can probably be made without recourse to a succession of meetings and discussions. If there are no more than two or three principals or directors agreements may be reached without much trouble. Nevertheless, a thorough investigation of the present position is necessary, and this should be set out on paper so that those concerned all start from the same basic information. Whoever is in charge of marketing should state how he sees demand in the future based on information he has gathered from the many sources available and his own judgement of how things are likely to develop. It should not be too difficult to decide the objectives over the next few years, and indeed to set the policy and the targets for the forthcoming year. The policy changes may be no more than decisions to try to get work a little further afield, or to aim to undertake types of work not formerly sought. The implications of these decisions on advertising policy, on staff, plant and transport requirements, should be worked out and the necessary action put in hand. Above all, the financial implications need careful thought.

In the case of medium-sized or large firms with more directors and more head office staff the procedures, although basically the same, are more difficult because more consultation is necessary.

Growth is best planned after considering all the factors involved. Only after everything has been carefully examined and discussed can a detailed plan be prepared. Every item which has been considered should be set down with a clear decision as to action required and by whom. A date should be fixed for a re-examination of the plan after the appropriate actions have been taken, but not necessarily before the results of these actions are clear, and periodical reviews should be undertaken to enable difficulties to be highlighted and remedial action put in hand.

Some of the matters that will need to be considered in the discussions leading to the determination of the plan are as follows:

(1) Is is proposed to keep only to the types of work previously undertaken? If so, growth can be achieved by seeking more opportunities to tender and this may require additional estimating staff.

(2) Is it intended to seek work in areas previously ignored? If so, this will mean more publicity in those areas and attempts to get on to approved lists of local authorities and other public bodies. It will mean making efforts to acquaint consultants and clients of the services the company can provide in these new areas. This may mean an increase in marketing staff as well as estimating staff. It will mean more expenditure on advertising and other promotional activities.

(3) Is it intended to seek contracts for types of work not previously undertaken? The marketing director will be able to advise on likely demand for various types of work. Consideration will have to be given as to whether the right kind of production staff are employed or should be sought, and whether additional or new types of plant are needed and, if so, should they be bought or hired? Much will depend on the company's beginnings. Did it start as a public works contractor or as a jobbing builder? Is it intended to concentrate most of its efforts on civil engineering or on building or to seek work in both categories? Is it civil engineering or building biased? Should there be additional staff engaged with the opposite bias to result in a more balanced staff? This is as important for estimating staff as for site staff. Building estimators generally

are not skilled in civil engineering estimating and vice versa. Similarly, civil engineering site staff are not always skilled in building and vice versa. In civil engineering much of the work is in the ground and is carried out using mechanical plant. In building the bulk of the work is above ground and interior finishes and installation of services such as tiling, decorating, plumbing, electric wiring, heating, ventilation usually account for a substantial portion of the total work. In building tolerances are more precise than in civil engineering and require more skill and more patience. The use of mechanical aids in building processes has been increasing at a steady pace for many years, but their contribution to output and manpower economy is a long way behind what has been achieved by the use of mechanical plant in civil engineering.

It may be that types of work previously carried out when there was an abundance of such work, e.g. opencast coal projects, are not to be pursued in future. Similarly, if the motorway programme shows signs of declining it may be decided that competition will be so keen that acceptable profit levels cannot be achieved. Both these examples raise a problem which contractors may have to face. It is the question of retention or otherwise of expensive heavy contractors' plant and plant operatives. Since lack of sufficient work in either of these fields will make it easy and less costly to hire plant if need be, consideration must be given to the questions of plant retention or replacement. Sentiment and vested interests must not be allowed to have too great a bearing on decisions on these important questions. There is little point in retaining plant for which the future use is very difficult to see, and this applies equally to plant operators.

Every aspect of what was formerly carried out by contractors' own employees, but is more usually now carried out by sub-contractors, must be examined realistically. Unless the work can be carried out more profitably by the company's own employees, and sometimes it can, consideration must be given seriously to reducing the sections of the company concerned. These will be difficult, even agonizing decisions, but there is little point in continuing with sections which cannot compete in price and sometimes in speed and quality with specialist or semi-specialist sub-contractors.

An extension of this line of thought leads to the consideration of the use of labour-only sub-contractors, a practice which is widespread. There are many very reputable labour-only sub-contractors and the practice is not new. It existed before the Second World War, particularly in house building. The important aspect of this practice is the cost of employing labour-only sub-contractors compared with using sub-contracting firms or a contractor's own employees. It is claimed that productivity is higher when labour-only sub-contractors are employed, but is this measured in terms of output per head, or output per week, or unit cost of output? It may, in some cases, be worthwhile paying more to get work completed more quickly, but this cannot be true for every contract. What about quality? Whether or not to employ labour-only sub-contractors is a production matter, but the decision affects cost, speed and quality and these are matters of importance in marketing considerations. The truth will be hard to find but a thorough examination is essential to the well-being of the company. The answer probably lies somewhere in the middle: in some cases use labour-only sub-contractors, in others do not, but make sure that the pros and cons have been properly considered before decisions are made.

Increased turnover may be achieved by any of the methods listed above or by any combination of them. One thing that must be said is that there is far more building work available than civil engineering and the best prospects for increasing turnover exist in the building field. This does not necessarily mean that civil engineering contractors must switch some of their activities to building, but it will be more difficult to expand in the civil engineering field alone.

When all the matters mentioned above have been carefully considered and decisions reached, action may be needed in several parts of the firm in the process of moving towards the chosen objective. The organizational structure of the company will need to be reviewed. Possibly some internal promotions of staff or some recruitment will be necessary. A new staff complement will be needed and the personnel manager or director or whoever is in charge of total staffing will need to set out the new staffing structure. A new organizational chart showing lines of control and responsibilities will be needed and this should be dealt with promptly, showing staff vacancies which need to be filled. Whoever is in charge of accommodation will need to carry out a review and, if necessary, to make changes in staff locations. Those in charge of plant and transport will need to make decisions about any increases or decreases needed, and to make suitable arrangements for adequate depots.

An alternative to this scientific approach is to try to achieve a new objective within the existing organization, with the assumption that all staff, or at any rate a good many staff, could undertake a little more work. This is probably true, and if a new objective can be achieved without any, or perhaps only a very few increases in staff, then staff costs in relation to turnover decrease, productivity increases and the firm becomes more profitable. There is a risk in this approach, but then there are always many risks in general contracting, so this is nothing new. The risk is that it will be found that work in some or other department begins to lag behind and the additional staff needed cannot be obtained quickly, even with a high level of unemployment. Agencies can be used as a temporary solution, but this should be avoided if possible.

Whichever method is chosen, the scientific method of deliberately trying to set up the best organization to achieve a new objective, or the pragmatic approach of not taking action until it is forced upon the firm by circumstances, will depend on the size of the firm and the size of the expansion proposed. In the smaller firms the pragmatic approach has something to commend it, but it could lead to a reduction in the quality of service to clients, particularly delays, and these must be avoided. In the larger firms the pre-planning method is better, and this is particularly so if the firm has out-stations where staff recruitment problems are unlikely to be uniform.

To summarize, it is suggested that it is necessary to take stock of the current organization and its turnover and profit, and then to decide what the objective is to be over the next five or more years. Once this has been settled, whatever changes are likely to be needed in the current organization should be assessed and action taken to set up the organizational changes necessary to achieve the new objective. Once again it is stressed that progress to the new objective is unlikely to be uniform, and patience will need to be exercised if initial progress does not come up to expectations.

Chapter 4

Marketing procedures and management

Before proceeding with some suggestions for the preparation of a marketing plan it will perhaps be appropriate to discuss broadly the type and structure of the management organization usually to be found in contracting organizations. Clearly these structures will vary according to the size of the enterprise, and it is not, therefore, possible to set out an organization which could apply in every case.

The basic requirements will, however, be found in one form or another in every company in the general contracting field whatever its size. These are estimating and tendering, production, buying or purchasing, surveying or measuring, and accounts. In small companies there will not always be separate individuals in charge of each of these functions. In the larger companies there may well be. The large companies may also have additional functions under the control of separate individuals and one would expect to see separate departments for sub-contract work, planning engineers, wages, personnel management, marketing, industrial relations, safety, plant and transport. The head office organization suggested in the box on page 32 would be suitable for a medium to large company.

The suggested marketing procedures and management which follow are directed mainly at medium-sized firms, although the general principles apply to firms of any size. The extent to which they can be applied will depend on the size of firm. The principal aim is to give a general indication of how marketing planning and management can be undertaken. The box on page 33 depicts a typical marketing organization in a medium to large company.

Director level	
Company Secretary	Insurance / Legal / Administration
Personnel	Staff / Labour Manager / Industrial relations
Finance	Chief Accountant / Purchasing / Data Processing Manager
Marketing	Marketing Manager – Company – Intelligence Company Photographer / PRO and Advertising
Estimating	Chief Estimator (Building) / Chief Estimator (Civil Engineering)
Chairman–Managing Director	
Surveying	Chief Surveyor – Site Surveyors
Production	Sub-contracts Manager / Planning Engineers – Contract Managers – Site Agents etc.
Technical	Technical / Design / Plant / Transport
Overseas	Organization depends on extent of involvement

The mechanics for formulating a marketing plan are essential to the production of a practical plan capable of achievement. It is suggested that responsibility for the first move should be placed squarely on one person. Ideally he would be the marketing director or manager or, if no one has either of those titles, possibly the head of estimating or quantity surveying should be given the task. He should prepare a paper for discussion which should have two parts. Part 1 should record the latest year's results in some detail. It should show turnover and profit by types of work and location. There is a difficulty here in that final accounts are frequently not settled until long after the projects are completed, but this can be overcome by including a column showing turnover and profit from projects completed in earlier years, and estimates for contracts still awaiting settlement. This, in itself, could be very illuminating.

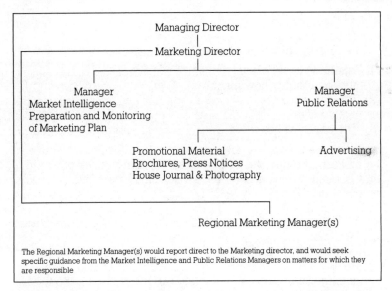

The Regional Marketing Manager(s) would report direct to the Marketing director, and would seek specific guidance from the Market Intelligence and Public Relations Managers on matters for which they are responsible

The production of a good comprehensive Part 1 is dependent on the existence of good records. It is far easier to assess what happened if record cards are readily available in one place. This saves time in consulting various departments about matters on which their memories may be vague.

Part 2 of the report should be an attempt to forecast the likely level of demand in the next few years. This will obviously be difficult to quantify, but that is not really necessary since there is always more work available in any sector than any single contracting firm can undertake, however big it may be. What is required is an indication of likely trends, e.g. more defence work of all kinds, more leisure facilities and so on. Guidance on trends is available in many places: the government publication on public expenditure, the published programmes of local authorities which appear annually in *The Surveyor*, the analysis of contracts and approvals published by *Contract Journal*, the articles on construction indicators in *Building*, the twice yearly reports by the NEDO for Building and Civil Engineering, an analysis of projects reported in the publications of Associated Building Industries and Salesleads published by Glenigan, both of which are available on subscription. The *Statistical Bulletins on Construction* published by HMSO are also useful and so are the monthly press handouts from the Department of the Environment. The total amount of information available is fearsome and requires hours of study before reasonable forecasts can be made, but it is not an impossible task, and must be undertaken.

The forecast of likely demand should state:

(1) The likely level of total demand – is it increasing or decreasing and by roughly how much?
(2) The types of work likely to be available and an indication of the likely values.
(3) The areas where these demands most probably will occur.

As has been said, it is not necessary to be absolutely accurate so far as Part 1 is concerned. Whatever the current level of output of any company, it can only be a very small percentage of total output. For example, in 1982 the output at current prices in the three sectors with which this book is concerned was (i) Public, £2927 million; (ii) Industrial, £2131 million; and (iii) Commercial, £3014 million.

A contractor working in all these fields, and most general contractors do, with an annual output of, say, £50 million, would be producing £50 million out of a total of £8072 million or about 0.6%.

With regard to Part 2, the important thing to realize is that there are likely to be opportunities for most types of work in the future and the aim should be to indicate which types are likely to be more plentiful and which less so. Part 2 should include contributions from the production, quantity surveying and estimating directors showing the capacity available. These should be expressed as total company capacity, but if there are regional or area organizations the capacities in these locations should also be indicated.

This paper, when completed, should be circulated to directors and should be discussed by the board. It should be the basis for decisions about future policy. If, as expected, the board is in favour of increasing turnover and profits, there are a number of matters which need careful consideration.

(1) What finance is available or can be obtained? Should further borrowing be undertaken? If so, what interest rates are likely and how are they likely to change?
(2) Is the current staffing level adequate, or are additions required and where? What are recruitment possibilities? Are some internal promotions needed? What additional costs would be involved?
(3) Are plant and transport levels adequate? Will new plant or transport be required and should it be purchased or hired? Should some plant be sold?

(4) Are the premises offices, workshops and yards adequate? Could they accommodate additional staff or plant and vehicles?
(5) If the firm has outstations, consideration should be given to possible effects on those parts of the organization.

These discussions should be undertaken by the full board with directors in charge of the various departments giving their considered views on the relevant aspects. This may mean several meetings, but it is essential if an attainable growth rate is to be settled.

It is suggested that the growth rate should be settled in terms of turnover at constant prices, and this should be expressed as, say, X% over the next 5, 10 or some other period of years. It can then be broken down simply as $(X/5)$% or $(X/10)$% per year for the next 5 or 10 years. This is obviously not strictly correct, but it is accurate enough for planning purposes. Unless constant prices related to a base year are used, it could be that a 5% increase on current prices represented a decrease if costs were rising by more than 5% per year.

If there is regular feedback of information within the firm, this will be of great use in forward planning. Indeed it will be possible to calculate planning factors for all manner of activities. Some of these are outlined below with an indication of how they can be used. Suppose the current turnover of the firm is £20 m and the planned growth is to attain £30 m at the end of 5 years, i.e. 10% per year on average. If work in progress still to be done together with new orders not yet started will result in output of £10 m in the next year, then new orders to the value of £24 m at constant prices will need to be obtained on the basis that the average duration of a contract is 12 months. This £24 m of new orders will produce output in the first year of £12 m and this together with the output from work in progress or ordered will produce the remaining £10 m to give an annual output of £22 m. These figures are not absolutely accurate and with the effects of keen competition and inflation, this aspect requires monitoring and adjustment as necessary.

On the assumption that £24 m of new orders are necessary, the tendering success rate should next be looked at. If this is, say, 1 in 8, then tenders for £192 m of work must be submitted. If requests to be included in tender lists has a success rate of, say, 1 in 4, then £768 m value of contracts must be pursued. This is not strictly correct, because some opportunities arise from former clients and their consultants without any pursuing action being necessary, but with about 100 contractors seeking every sizeable job up to about

£2 m the success rate so far as getting opportunities may not be quite so good as 1 in 4. On the other hand, as jobs get bigger, the field of contractors anxious or able to undertake them gets smaller and the success rate improves greatly. In the case of projects between £500 000 and £2 m there is very keen competition for places in the tender lists, and there is a great deal of hard work necessary to get a satisfactory proportion of the enquiries.

Further planning factors based on feedback will help to give guidance as to staff numbers needed in various departments. For example, if it is necessary to quote for work to the value of £192 m to obtain the desired value of successful tender, the use of planning factors will indicate the necessary size of estimating departments. The output in the past of a civil engineering estimator may be about £20 m per year, i.e. one estimator and, say, four colleagues can cope with £100 m of estimating in a year. For building estimators, the planning factor is higher because there are more prime cost items in the bill of quantities. It may be of the order of £30 m per year. Applying these factors to the civil engineering and building components of £192 m, say 40% of civil engineering, 60% building, the staff requirements are roughly four civil engineering and four building estimators. Of course, with prime cost items and with other work being sub-contracted by decision of the contractor, these figures may need modification, but they are of the right order.

In theory, there is no end to the use of planning factors. They can give guidance to the number of agents, general foremen, surveyors, shorthand typists, switchboard staff, accounts staff and to the amounts of plant and transport needed. They can also give guidance as to the area of office space needed for a given number of staff.

Without doubt, matters of the kind mentioned above are dealt with in small and even in large firms by instinct. How far these are accurate has been demonstrated to some extent by the redundancies which have occurred in many firms. Whilst the use of planning factors is not an exact science, some guidance based on feedback is better than guesses based on impressions unsupported by facts.

Job records

It has been suggested above that adequate job records should be kept so that proper consideration can be made of all manner of

Region	Name of project and classification*	Client	Consultant	Qty. surveyor	Form of contract
Tender No.	Director in charge	Contract manager	Project manager/ Agent	General forman/ Engineer	Surveyor
Value	Date started	Contract completion date	Actual completion date	Final account and date	Profit/loss
Notes					

* Classification of contract should be selected from a list of types of work used throughout the company, e.g. factory, warehouse, office, school, road, motorway, bridge, hospital. A code number can be used for each type and the list can be as long as considered necessary.

aspects in drawing up a report on past performance and in making proposals for the future. What form should these records take, who should enter the information and where should the records be kept? A simple record card for each project is shown on page 37. It is not claimed that this is perfect, but it shows information which should be readily available to anyone in the organization. To this end, it is suggested that the cards should be kept in the marketing department because they are likely to be the most frequent users of them. There may be a case for copies to be kept elsewhere, but it is better if there is a central information office where anyone can find out what he requires about any contract.

The cards should be originated in the estimating department and passed to the production department for the details of staffing and starting and completion dates to be entered. The cards should then be passed to the marketing department for filing in a Kardex or similar system. The actual completion date should be notified by the contracts manager to marketing as should the final account settlement and the finance department should notify marketing of the profit/loss. Space should be left on the card for any special comments which become necessary. These cards would provide full information about the job and would help marketing in their promotional efforts.

An important consideration affecting the production of a marketing plan is when should the preliminary work be undertaken? Clearly, there is a good deal of research necessary, and departmental discussions as well as those at board level will all take time. If it is hoped to operate a marketing plan from, say 1 April, it is clear that the research work and the discussions must take place and the final plan drawn up some time in advance of that date. The initial work will need to start in the previous year and possibly not later than June. This should give sufficient time for all the thought processes to take place and for firm decisions to be reached early enough for the plan to commence on 1 April. There is a difficulty in that in the previous June the performance of the company over a full year will, at best, be for the year ended 31 March. With only half of the current year gone, the performance in that year will not be known, and only an estimate of likely performance will be available. This may be good enough to take as the base year, but if not, the previous year should be taken as the base year for the first marketing plan. The plan can be adjusted if necessary as more accurate estimates of the current year's performance become available with the passage of time.

The existence of regional offices or other outstations creates other problems which can be solved if proper record cards are kept, as described earlier, and if programmes and performance are recorded in a form similar to that suggested in this chapter.

Companies that already have a marketing plan will have given careful consideration to the basic information needed and how it is to be recorded before decisions can be taken. Experience will have enabled them to deal with difficulties which have arisen and the annual review of the plan, and indeed its regular monitoring, will have become routine. Those companies that have not yet decided to work with a marketing plan but are considering whether they should do so may find the suggestions in this chapter helpful. If must be said that there are many ways in which this planning work can be undertaken and individual firms will, no doubt, experiment until they find the way that suits them best.

It is suggested that programme/performance charts will be helpful in the monitoring of a marketing plan, and a method of drawing up such charts is given in Table 4.1.

Table 4.1 First year planned programme 1984–5

	Contract value (£)	Output to date (£)	Planned output 1984–5 (£)	Output to complete (£)
Projects under construction on 1/4/84				
1. Factory, Telford	641 000	220 000	421 000	–
2. Sewage Works, Warwick	863 000	150 000	650 000	63 000
Orders received but not started on 1/4/84				
1. School, Walsall	610 000	–	470 000	140 000
2. Road, Stafford	879 000	–	879 000	
New Orders (Total required £9 m)				
1.				
2.				

New orders should be entered as received and outputs estimated and entered in the appropriate columns. This document should be prepared by the marketing department in conjunction with the production department, and copies sent to all directors each month end. Amendments to the estimated outputs should be made as necessary so that the document gives a clear, up-to-date picture. Similar charts should be kept in outstations and copies sent to the head office marketing department each month.

From the beginning of April onwards the marketing director will undertake continuous monitoring of the plan, bringing to the attention of the managing director any matters that need action. Monthly board meetings will receive progress reports and suggestions for amending the plan as and when necessary.

The first year's results are important and should be reviewed by the board as soon after the year-end as possible. There should be no alarm if the results do not match precisely what was planned. It would be surprising if they did, but they should show progress towards the long-term goal.

Because it is not advocated that the long-term plan should be over-ambitious, the annual growth looked for in the case of medium or large firms, already having a sizeable turnover, is likely to be relatively small, perhaps no more than 5 or 10%, and this should not be too difficult to achieve using the methods described above. In firms with smaller turnovers, say from £200 000 to £1 million per annum, planned increases of 100% in five years are attainable if all the conditions are right, i.e. if there are no serious problems about finance, staff, or accommodation and if the firm is truly competitive. This sort of growth approximates to about 20% per year and does not look excessive. It is largely a question of determination coupled with sensible pricing and good service.

Although it is not proposed to consider estimating processes, which are a specialist function best left to the estimating department, the marketing director will clearly be interested in the success or failure of tenders submitted. His advice on which competitors will be particularly keen to obtain a contract under consideration will be of help in the final decision on the price to be submitted. Tender prices should not be submitted which contain no element of profit and that means that overheads and site costs must be kept under continuous review. These can always be reduced if the will is there, and if the problem of vested interests can be overcome. Once again, the importance of a proper staff costing system and feedback is stressed.

Whatever the size of a firm, long-term planning must be undertaken at head office. This is so obvious as to be hardly worth mentioning, but it is mentioned because in some firms where there are regional organizations or other outstations which are not separately registered companies there is a tenency to allow these outstations to determine their own plans or budgets. Whilst the outstations need to be brought into the total planning considerations, it would be wrong to allow them complete self-determination. They are part of the firm and their efforts need not only head office direction but head office monitoring on, say, a monthly basis. Outstations must not be allowed to determine the policy of the firm, although their views need careful consideration, and therefore they cannot be allowed to say what contribution they will make to the total plan. They must be told what contribution they are required to make, and given authority to acquire the resources they need to achieve their contribution. It goes without saying that in formulating the draft plan for the company as a whole, proper consideration of outstation contributions in the past and possible increases in the future must be assessed in consultation with outstation managers. Indeed, the starting point is probably for outstation managers to set out their current or immediate past achievements and staffing and to suggest what they feel they could achieve if required and also to state what changes in organization, staffing or procedures would best help to that end. For example, would it help if they were given full authority to prepare and submit quotations instead of having to consult head office in every case or at least in some cases? Would it help if they could deal exclusively with their staffing problems or would that lead to overstaffing of the company? Should they be free to do their own purchasing, or would this lead to less favourable prices for materials than if there were central purchasing? Delegation is the only logical way to run big organizations, but outstations are only part of an organization and they cannot be given a completely free hand. Would it be sensible to give them working capital and allow them to manage the financial aspects of their work? Would it be sensible to allow them to buy and perhaps sell plant or to decide when to use company plant and when to hire? If they are to be held responsible for their success or failure, they must be given the necessary authority to manage with recourse to head office advice as and when they need it, but not to be under day-to-day control from head office. If the outstation manager is not allowed to manage his organization in the way he believes to be best, he will never be able to manage anything. The maximum delegation seems most sensible within a policy clearly laid down by head office. So the outstations must help in preparing the overall plan, but be told in the end what their planning target is. They then inform their senior managers and their senior staff representatives in the same way as this has been done in head office. A suggested regional organization for a medium to large company is shown in the box on page 42.

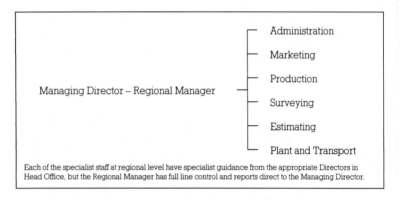

Each of the specialist staff at regional level have specialist guidance from the appropriate Directors in Head Office, but the Regional Manager has full line control and reports direct to the Managing Director.

Outstation advice will be particularly useful about types and locations of work. For example, it may be perfectly sensible for head office to undertake major motorway contracts, but not very sensible or even downright folly for regional organizations to do so. The reasons are fairly obvious: lack of experience, lack of suitable plant, lack of suitable staff and operatives. The trouble with this sort of policy is that the company will then tend to undertake this type of contract only within daily travelling distance of its head office. By doing so, it reduces the opportunities available.

The logic of the above example is that there should be a relatively small head office organization and that working organizations wherever they are should all be on the same basis charged with responsibility for carrying out work in a particular area. The idea that head office can perform the dual role of policy making and yet have operational responsibilities for the area around the head office leads to policy being based on experience of operations only in the area covered by head office, and this often produces nonsenses so far as operations in other areas are concerned. Is it a marketing function to make such comments about organizational matters? It most certainly is, if the Institute of Marketing definition set out in chapter 1 is accepted.

It may be considered in preparing the overall plan that there are good opportunities for profitable contracts in areas of the country which the company has so far ignored. If this view is held, there may be some who would advocate opening a new office in that area with the intention of seeking and securing contracts. This is not a line to be pursued. It might take many months before a suitable opportunity presents itself and even that might not lead to a contract. A much better way is to seek opportunities from the nearest existing office (outpost or merely site office address could be used). In this way no new organization is eating its head off whilst waiting for the first contract. Once a contract is obtained, the company is operating in the new area, its site boards are visible and so are its transport and plant. Its personnel gradually get to know people in the locality, consultants suppliers, sub-contractors and this makes it easier to get further opportunities leading, hopefully to contract no. 2. With a third contract in the area, the company is established there and can then open a small office for further development in the area if the prospects are favourable.

There is another method of getting a foothold in a new area, namely to acquire a local firm there. There are often firms willing to sell and sometimes firms with a high reputation. Most big firms have acquired others during their growth periods, sometimes because of an expertise lacking in the purchaser, sometimes merely to obtain a base in a particular area where opportunities were regarded as likely to be available for further development. This method of expanding is most useful and should not be ignored.

It goes without saying that the most careful examination of the firm's reputation, finances, staff, plant and accommmodation is essential before any firm commitments to acquire are made. Similarly, firms specializing in particular types of work can be acquired to fill gaps in ability or to strengthen weaknesses. For example, it may be decided that a firm specializing in piling or tunnelling or in road surfacing would be a useful addition. Consideration would then have to be given as to whether such an acquisition should operate as a subsidiary or whether it should be absorbed into the existing organization. Much would depend on whether the purchaser could use all the capacity of the acquired firm itself or whether it would wish to quote to other general contractors. Another consideration is whether it would be advantageous to retain the acquired company's name which will, perhaps, be well known in its locality or even nationally, and, of course, its senior staff who would, no doubt, have useful contacts. What is not recommended is the total replacement of the senior staff by people from the acquiring company, although consideration should be given to strengthening weak spots in the acquired firm's organization. Clearly, it will be necessary to introduce into the acquired firm the policy and procedures of the purchaser. It seems obvious, therefore, that the manager of the acquired firm must be replaced at once by someone from the acquiring firm who is experienced and familiar with the policy and procedures.

Chapter 5

Promoting the company through advertising

No one knows the true worth of an individual or a company unless someone has drawn attention to it. Actions may speak louder than words, but the words are needed to identify the person or firm behind the actions. Much excellent work by individuals and companies has gone unnoticed because no one took the trouble to publicize it. Self or company promotion is essential if proper rewards are to be gained. In the case of individuals it is considered to be a form of boasting – blowing one's own trumpet – if one is for ever drawing attention to one's successes – or is it? Is this why so many people in the entertainment world employ agents to secure engagements for them? Is it that they employ a trumpeter instead of blowing their own, and is that not called boasting?

This book is not about individuals; it is about contracting companies and how they can improve their profits by adopting marketing principles. One of these principles relates to promotion of the company: to getting the company's ability and achievements, the resources at its disposal and the services it offers known to prospective purchasers of those services.

There are many ways in which this can be done and one has only to watch commercials on television, to read the national and local press, to see the hoardings and neon lights, and to see the delivery of unsolicited trade material which daily pours into offices and residences to appreciate that advertising is one of the main methods of promoting a business undertaking.

In 1979 expenditure on advertising in the UK in the media, excluding that on outdoor signs and at exhibitions, totalled £2129 million, or about £380 per head for every member of the population. This is a colossal amount. Since 1960 expenditure has increased in real terms by about 25%. Nearly two-thirds of this was for advertising in the press, including magazines and trade journals, and about one-fifth on television. Again, nearly two-thirds was spent in display advertising in the press. The biggest spenders were various well-known stores and very few manufacturers appeared in the Top Twenty and, of course, no contractors. This is not difficult to understand. Except in the case of house-builders who have finished products for sale which can be inspected before purchase, contractors have on offer only services. This does not mean that advertising by contractors is unnecessary. There is plenty of evidence that contractors do advertise and that some, at least, spend considerable amounts on advertising in all manner of places and by various methods.

There is a need for contractors to advertise just as any other business undertaking needs to inform prospective purchasers what they have to offer. Advertising for contractors arises in many ways which will now be examined in detail.

Corporate advertising

Corporate advertising is a means of communicating to the public at least, or a section of the public, the nature of a business undertaking and the services it offers. It helps prospective clients to understand the company's aims and objectives. Corporate advertising is all about creating an identity or an image of the company and of selling the company to people who may one day be able to contribute by way of orders to the company's growth. The company can decide how it wants the general public, the man in the street and, above all, the decision-makers in industry, commerce and public authorities to regard it. It can choose whether it wants to stress its public spiritedness, its determination to help improve the quality of life, its financial soundness, or its technical ability. It can decide whether it wants to demonstrate its ability to deliver what is needed on time and at the proper price.

Not only is corporate advertising a means of creating the desirable image to decision-makers, but to other people of importance to the company: to the Stock Exchange for public limited companies, to shareholders, to the institutions, to the financial editors of the press who can influence opinion so easily and to trade union leaders who, these days, have so much influence in many directions.

Quite apart from these external people who need the clearest picture of a company, corporate advertising has a pronounced effect on the company's own staff and on possible future staff. In recruiting the next crop of engineering or building graduates, nothing is more galling than to be asked, 'Who are you and what do you do and where?' Existing staff need to be kept informed of policies and plans and they will all see corporate advertisements and make their own judgements as to whether the image created is in line with their own views.

In considering policy on corporate advertising, there are three things which need decisions:

> What is the message to be?
> Where shall it be given?
> How frequently?

What should advertisements of this nature say? What is regarded as the real strength of the company? Is it its financial soundness? Is it its utter reliability to carry through successfully whatever it undertakes? Is it its contribution to a better quality of life? These and other aspects need careful consideration and the advice of an expert is essential. It is not suggested that there is only one message to be transmitted. The occasions and the media chosen, not to mention the general policy of the company and the current or foreseeable state of trade all have a bearing on the choice of message. It is regarded as essential to employ an advertising agent by all but the smallest companies. Their function and cost will be discussed later.

One thing is certain: there will be no lack of opportunities for corporate advertising. Many of the national and regional daily newspapers present frequent opportunities for this kind of advertising. Well favoured positions very much in demand are those corner pieces – sometimes called title corners or earpieces – at the top of the first page or the business page of the well-known quality newspapers whose readership is drawn mainly from politicians, senior central and local government staff, and those holding important posts in public and private enterprises. These are the people who have the purchasing power for construction work, but are not necessarily the people who actually decide which firms should be given opportunities to quote for construction contracts.

The space available in these cornerpieces is small – no more than about 2 × 2 inches. The message which can be conveyed is, therefore, restricted by the space available, and this means usually that little more than the name, address and telephone number of the company and a simple statement of its function can be included. The demand for these spaces is very great and once secured they should not be given up. Their usefulness lies in their being on the front page or the first page of the business section of the newspaper. They cannot miss being seen by readers and their regular appearance will certainly mean that the company's name and function are firmly imprinted in the minds of those who see them. They must, therefore, appear at fairly regular intervals, and it is necessary to take a series of 12, 10, 8 or 6 a year, all exactly the same.

The principal job they do is to keep the company's name and function before the decision-makers who are likely to be the readers of the quality newspapers. A good deal of thought should be given as to the precise copy to be used and in which of the many newspapers the advertisements should appear. The company's advertising agent will be able to advise on both the copy and the media, and will have the circulation figures for all the newspapers.

In the next chapter on public relations there is a section dealing with company logo type, i.e. a symbol which it is hoped will be instantly recognized as the sign of the company. There are a number of the largest companies in the industry that have achieved this, and there are many that have not yet reached that level of recognition. If the company has devised a suitable logo type, it should be used in all its advertisements, on its letterheads and on vehicles and plant whenever possible.

Annual advertising

In drawing up an advertising programme or budget, it will become apparent that there are a great many annual publications of interest to all concerned with construction. These include the well-known directories, the year-books of the various professional organizations such as RIBA, RICS, Institute of Structural Engineers, Association of Consulting Engineers, Concrete Society, NFBTE and its constituent associations, FCEC, FMB and others. Although there is some doubt as to how frequently these publications are looked at after the initial publication, they are, nevertheless, the guide books to the various associations. Similarly, many public authorities and particularly local authorities, are concerned with the publication of books devoted to particular aspects of their work. For example, some local authorities have handbooks dealing with the work of their planning department, or education, or amenities. Some have town guides. There are throughout the country literally hundreds, perhaps thousands of these publications. They cost the local authority no more than the staff time involved in preparing the editorial matter because the publishing company recoups the cost of production and distribution by selling advertising space. The prime mover in almost every case is the publisher and local authority staffs see a relatively cheap (to them) way of informing the public of their work and procedures. These publications are full of information, useful and sometimes essential to those in the construction industry and are, therefore, suitable media for advertising. But the number of these publications and those of professional associations is so great that it is quite out of the question to take space in all of them. It is advisable to take space in a selection, not necessarily every year. By alternating year by year cost can be kept within reasonable limits and the company's name and function spread over a wider readership. Again, the advertising agent's advice will be helpful.

Other opportunities for advertising occur in the diaries issued annually by police, fire brigades and other authorities. Police and fire brigades are likely at some time to be of assistance to companies in the industry and there is much to be said for taking small advertisements in their diaries. The cost is usually very small, and it is more an expression of support than the hope that it will bring in any business – but it might.

Series advertising

In addition to the earpieces referred to above, there are endless opportunities open to contractors for advertising in magazines, journals, the trade press and other daily, weekly or quarterly publications. Some are much better vehicles than others for company advertising. For example, there is not a great deal of point in advertising in publications whose principal readers are one's competitors and their employees. It is much more advantageous to advertise in publications mainly read by those in a position to influence the placing of orders, although it must be said that advertising in *any* publication might catch the eye of a likely future client. The availability of hundreds of publications does nothing to make the choice of media any simpler, but obviously the placing of advertisements in journals of the professions concerned with construction are more likely to result in enquiries than others. This does not mean that the journals of the RIBA, RICS, ICE, should be the only ones chosen, but these appeal straightaway because of their readership.

Again, there is not much point in a single advertisement in any of these publications. This would make very little impact, and a series of identical advertisements is a much better course. There are reduced rates for a series of various lengths and how frequently copy should be placed will depend to some extent on the size of the advertising budget. The advice of the advertising agent will be helpful.

There is not a great deal of point in advertising in this type of journal which has a national readership unless the firm operates on a national, or nearly national, basis. For a firm that operates only in a specific locality advertising in national publications is a waste of money. There are other local vehicles which are better for those firms, and local and regional newspapers are the obvious ones. Some development corporations and other public authorities have journals published monthly or quarterly which include advertising opportunities and these can be useful for the more localised firms.

The type of advertisement, that is the copy, to be used in serial advertising needs careful study. It may be considered that a corporate type of advertisement can be used effectively, or a more specific one dealing with some aspect of the firm's search for work is more appropriate. For example, the copy for an advertisement in a journal mainly read by civil engineers should be different from that used in one read mainly by architects. There is little point in telling civil engineers that the firm has a good record of hospital building, or of telling architects of the firm's success in constructing sewage works or roads.

What form the copy will take will depend on the journal chosen, the types of work sought and the readership. It is neither essential nor desirable to cram the copy with details of every service offered, but there is much to be said for giving a few examples, carefully chosen, of projects successfully undertaken. It is surprising that few people even in the industry can name the design architect or engineer of many important buildings or civil engineering projects completed in the last few years. Even fewer can name the main contractor. It is, therefore, likely to be beneficial to tell prospective purchasers and their consultants some of the projects the firm has built, but not too many in one advertisement. It is also helpful to include one or two under construction.

Although advertising copy is not always read completely by readers of journals, if it is eye-catching enough at least somebody will read it. Too much print is to be avoided. Something short, easy to read and bringing out any special aspect of the firms' capabilities should be the objective.

The use of photographs in advertising copy needs to be carefully considered. They certainly catch the eye and, used skilfully, can be most effective. There are three main difficulties in their use. First, they would almost certainly have to be black and white, although colour is often possible. The use of colour is more expensive and it is doubtful whether it is worthwhile in these types of advertisements. Second, reproduction of black and white photographs on newsprint is not very good and colour even worse and for this reason the use of small photographs may result in a very disappointing advertisement. If photographs are to be used, the size of the print needs to be fairly large. It should carry a caption saying what it is and where, and if possible the name of the client and the consultants should be stated. Third, care must be taken to ensure that the photograph does not show anything that might be detrimental to the firm. This is particularly so in the case of projects under construction, where untidy sites and other things which might lead to some criticism should be avoided.

Before photographs are used in advertisements, it is essential to ensure that the client has no objections. The necessary clearance can usually be obtained from the consultant and it is preferable to consult him rather than the client direct. Some bills of quantities specifically debar publicity and this must be honoured.

Some firms use drawings in their advertisements. These are often in the form of a composite picture showing various aspects of construction. If this type of advertisement is favoured, the freehand sketch will probably be made by a commercial artist employed in house or commissioned by the advertising agent. There are many capable artists available, but their ability is in sketching, and they have not usually had experience of working in a construction company. It is, therefore, essential to examine such a sketch before passing it for publication to ensure that there are no constructional nonsenses in it, e.g. a type of plant depicted in use on an aspect of construction where such plant is not at all appropriate, or workmen not wearing protective headgear, or no toe-boards on scaffolding. Experienced members of the firm can spot these errors very quickly, and so will those who read the advertisement if they are allowed to go through without correction.

It is said above that the choice of media should be influenced by the readership of journals most likely to be in a position to place orders. This does not mean that trade journals of other industries should be neglected. All industries require construction work at some time, and so their journals can be good vehicles for advertising. For example, the pharmaceutical industry is always very active and if a firm has successfully undertaken, say, a production or a research establishment for a well-known firm in this industry, it will do no harm to advertise the fact in the pharmaceutical journals. This applies to many industries. The advertising agent will be able to advise on the most likely journals to be considered and, again, a single advertisement is not likely to be very effecive.

How to judge the effectiveness of advertising

Advertising a construction service is very different from advertising a finished product. A prospective purchaser of, say, a motor car may have been attracted to a particular make or model as a result of an advertisement. He can go into a showroom and see the car. He can examine its lines, its finish, he can sit in it, he can drive it, he can enquire about petrol consumption and servicing and the price of spares, all before he makes any move towards purchasing. Something similar is available for prospective purchasers of most capital or consumer goods, particularly the latter. There are opportunities for detailed inspection of the goods before purchase and in many cases a money-back guarantee if the purchaser is dissatisfied after purchase. This is not the case in general contracting in the construction industry. The purchaser cannot pre-examine the product. The contract is placed on trust.

It follows that advertising for most capital and consumer goods is very different from advertising a service, and the effectiveness of advertising is more easily measured in those industries than it is in construction. In consumer goods industries the effect of a television or newspaper advertisement can sometimes be seen the next day by an increase in prospective customers in the shops. For construction no such thing happens. Indeed, it may be months or even years before anything results which can be tied to a particular advertisement or there may never be such evidence. Advertising over a long period may persuade purchasers or their consultants to give a firm an opportunity to quote, but it is always difficult to tie this back to advertising.

There are ways in which contractors can attempt to measure the effectiveness of their advertising. The simplest is to include a coupon in the advertisement, or at least a statement, saying further information or literature can be obtained by ringing Mr X on telephone number so and so or by writing to him at the address given. This method is sometimes effective, but not all those who get in touch are prospective customers – at least not at the time they make contact, although they may be some day. Many of the replies come from students of construction and this is no bad thing and many come from firms wishing to sell to the advertiser!

Another method is to carry out market research. This can be useful, but it is expensive. A market research company will undertake a survey amongst purchasers and consultants on a sample basis and prepare a report. This report will show:

(1) What percentage of those approached had ever heard of the firm;
(2) What percentage knew what its business was;
(3) What percentage had ever seen an advertisement;
(4) Which publications these had been in; and
(5) How many knew of a project or projects carried out by the firm.

These market research companies are skilled in devising the right questions to ask and in analysing the replies and this sort of exercise is commended to firms that are in doubt about the effectiveness or the value of their advertising.

Display advertising

There are many opportunities each year for advertising in special features published in newapapers and journals regularly or to mark some specific event. For example, the heavy newspapers have features annually or quarterly on building and/or civil engineering, or on a region or Wales or Scotland, on the economic situation, or on training or marketing, or on industry in general, or on the progress or completion of a construction project of national or international importance. It is not posible nor is it necessary to attempt to give a comprehensive list of sorts of features that are published for they cover very many topics. The point is that whatever the topic, the construction industry is, to some extent, involved or could be involved. For example, a feature on education nationally must deal with educational buildings, their inadequacies or adequacies, their needs in terms of numbers, sizes, locations, design and costs. Similarly, a feature on banking might have only a passing reference to buildings for use by banks, but since contractors and investors in construction rely conspicuously on banks for financial arrangements on a day-to-day or long-term basis, such a feature is of some interest to contractors.

A case could be made out for contractors to take advertising space in any such feature, and certainly the publishers will approach contractors in their attempts to sell advertising space to support these features. It would, however, be out of the question for any contractor, even the very largest, to contemplate such a course and so a selection must be made, even if the end-result is to take space in none of them. A decision to stay out of all such features for large and medium-sized contractors would be both unusual and unwise. Some of these features present good opportunities for display advertising which should not be overlooked if only because a company's principal competitors will be taking space in them. This may have a ring of 'keeping up with the Joneses' and, if so, then it must be understood that if one does not keep up with the Joneses, one falls behind them.

It is suggested that the feature programmes, usually for a year ahead, published by the leading daily papers should be studied and decisions made as to which features are the best for contractors' advertisements. Is it better to take space in a feature that will be supported by one's competitors, such as a building and/or civil engineering review, most likely to be read with some care by those concerned with construction, or is it better to take space in a feature on, say, industry in general, or Wales, not specifically devoted to construction, which will be carefully studied by readers from a much broader field of activity? This is the sort of question which will need to be answered before a list of display advertisements to be taken can be determined.

What form should display advertising take? Whilst the type of corporate advertisement dealt with earlier might well be appropriate in some cases, consideration should be given to a more specific form designed especially to line up more adequately with the feature. For example, if the feature is to be about enterprise zones, the advertisement should address itself to the way the company can be of assistance in the development of those zones. If it is about hospitals, the advertisement should be about the experience and resources of the company to deal with hospital building and might contain examples of hospital projects successfully undertaken by the company.

In considering display advertising in connection with special features the advice of the advertising agent will be invaluable. He will be able to draw on his long experience of this sort of advertising and to advise on the best vehicles and the best copy. The important thing here, as in all advertising, is to try to ensure that the advertisement is seen and read and is not just another space very similar to many other advertisements. Contractors tend to be very conservative, and this is reflected in much of the advertising they place. It is not suggested that some of the more outrageous eye-catching advertisements should be copied, but there is scope for a great deal of improvement in copy and a well-chosen phrase which might become something of a household word is worth a good deal of thought.

Support advertising

The award of a contract frequently leads, either immediately or at some stage of construction, to a wish by the client to obtain publicity for the project. In such cases, the contractor is often, but not always, given publicity by the client in one way or another. In almost all such cases, there are opportunities for the contractor to obtain publicity.

These opportunities occur in both the publicly and privately financed fields, but more frequently in the latter. In the case of the former, when some important contract is placed, e.g. a motorway section, new shopping areas, large new offices, the authority frequently issues a press statement, and this is reproduced wholly or in part by the daily or weekly press and by the trade journals. It is important that the press handout mentions the name of the contractor and this is usually the case. There is not normally a special feature in the press on such occasions and so the possibility of taking advertising space does not usually arise, but when it does space should be taken.

In the case of privately financed projects, the client sometimes asks particular newspapers or journals to include a feature article about the placing of the contract. They do this because they consider it an occasion for obtaining publicity. This is particularly so when the project, when completed, will provide manufacturing or distribution facilities for the client's products or services, and the client's aim is to bring the development to the notice of potential customers.

Where such features are to be published, the editorial matter usually refers to the eventual use to be made of the facilities. Sometimes the name of the contractor who will build the project is not mentioned. It is advisable for the contractor's name at least to appear in the editorial and this is usually not difficult to arrange with the client. Some brief details of the contractor and possibly which members of the contractor's team will be responsible for the construction work, together with the start and completion dates is all that is required. Clients are customarily very co-operative in this respect.

The newspaper or journal in which the feature is to appear are anxious to sell as much advertising space as they consider suitable to be associated with the feature. They will certainly solicit the contractor, sub-contractors and suppliers to take space. It is somewhat churlish not to take advantage of such opportunities, but there is no need for lavish expenditure. All that need be said is that the construction work will be carried out by the contractor. The most effective way of doing this is to take a strip advertisement, about 8–10 cm deep, full page width, across the bottom of the feature page something like this:

(company logo)

Main Contractor for the Supermarket for Slick PLC at Wigginston

J. D. GARLESTON PLC
145 Hermitage Road
Hemiston

Tel. Hemiston 2463

Similar features may be requested by the client on completion of the project or when it is open for business, and advertising should be dealt with in a similar way.

Even publicly financed projects sometimes request features when official openings are to take place and again advertising should be contemplated as suggested. There have been cases where features have been requested by public authorities on completion of construction work and well in advance of official openings. For example, new hospitals prefer publicity on building completion because the official opening may be considerably later when beds and other equipment has been installed and when staff have been recruited. Obviously, a hospital cannot start to operate until it is properly furnished, equipped and staffed, and publicity at building completion stage is useful, especially for recruitment of staff.

There is another occasion when the opportunity for advertising arises. This is when the building has reached the stage where a topping-out ceremony might be appropriate. In this case, the initiative should usually come from the contractor, although occasionally the client makes the suggestion first. The procedure so far as advertising is as described above, but on these occasions, the editorial might well be prepared by the contractor.

These ceremonies will be dealt with in more detail in chapter 6.

Advertising budgets

There are countless opportunities for advertising in the press, on radio, televsion, hoardings, sporting events and so on. It would be easy to spend very considerable sums, but contractors generally do not begin to match the millions of pounds spent annually by some companies in the manufacturing and distributive industries.

Some attempt must be made by a contractor to arrive at a sensible budget for all its advertising. This amount may be derived in a number of ways:

(1) *An arbitrary amount* with no more guidance than perhaps, 'Let us see how things go.' This may be all right where advertising is not recognized as a definite and important part of company promotion, but in practice it has very little to recommend it.
(2) *A percentage of the previous year's turnover or profit.* This method disregards plans for the coming year and may restrict the amount of advertising really necessary. In today's circumstances, unless allowance is made for inflation, such a formula will lead to a reduction in real expenditure on advertising.
(3) *A percentage of the planned turnover or profit.* This is a more sensible method. On the assumption that the company would plan to increase turnover and profit, this will lead to increased

advertising expenditure and this in turn may lead to even greater turnover. Again, inflation must be taken into account.

If the planned turnover is to be achieved in part by seeking work in areas hitherto neglected or by offering services not previously offered, e.g. contract management, or prime cost plus fee, then additional advertising will be necessary and account must be taken of these proposals.

(4) *An amount based on an advertising plan prepared in detail, and subject to scrutiny by the Board.* This method is the most logical and can take account of all the relevant factors. Proposed space purchasing should be listed in detail with the cost under the following headings:

(a) Corporate Advertising;
(b) Annual Advertising;
(c) Series Advertising;
(d) Display Advertising;
(e) Support Advertising;
(f) Miscellaneous.

In such estimates a contingency sum should be included because there will undoubtedly be advertising opportunities which could not reasonably have been foreseen and which ought not to be missed.

When the estimates have been prepared, the marketing director should submit them to the board for approval or amendment. Any changes requested in the total sum should then be the subject of a further examination to see where the changes can best be accommodated.

Other advertising opportunities

In addition to advertising in the press, including newspapers and journals, and in year-books and directories, there are many other opportunities, but whether they are cost effective needs careful consideration.

(1) Radio

Now that there are many commercial radio stations which broadcast almost continuously, there are opportunities for advertising for contractors. Although these broadcasts have wide coverage, expenditure on radio advertising is not worthwhile. There are other methods of obtaining publicity through local radio at no cost which will be dealt with later.

(2) Television

Over 90% of the population can receive independent television programmes and so coverage is enormous. Cost is, however, high for even a very short advertisement, and advertising on television is not really suitable for general contractors, although it has been used by house builders. The latter have products for sale, and television advertising is probably best for those with products for sale. However, it must be said that some service industries have spent very large sums on television advertising.

(3) **Hoardings and other outdoor signs**

There are countless opportunities available for advertising on static and mobile sites.

The most obvious ones are the firm's own boards on its construction sites. Quite apart from the usually large boards listing the nature of the project, the client and the consultants as well as the contractor, frequently included as a contract requirement, there are always opportunities for erecting a simple board showing the contractor's name and logo. No opportunity for erecting these sign boards should be missed. The siting of them is important. They should be erected so that they are seen by as many people as possible. Facing main roads, on bridges, facing railway tracks, hung on to a building, on tower cranes are prime examples. Contract managers should make a point of checking that the best positions are used for this purpose. It goes without saying that the boards should be clean and sound. On conclusion of a contract, the boards should be returned to the company's depot for inspection and, if necessary, restoration undertaken before being issued to another contract.

The company's own transport and plant should also be used to publicize the company. It is a good idea to determine a standard colour or colours for all the company's transport and plant, and the choice of colour is important. It should, if possible, be an uncommon colour, one unlikely to be used in precisely the same way by a competitor. It may be that two colours can be used in an effective way to make a unique appearance. The vehicles and plant should clearly show the company's name and logo, and the best place to do this is on the back of the vehicle where it will be seen most. It should also be shown on the sides of the vehicle where there is usually more room, and on the front. These vehicles, moving around as they do, have a very useful part to play in company publicity. Again, the vehicles and plant should be cleaned regularly, and not allowed to accumulate mud and other dirt which not only gives a bad impression, but can obscure the company's name.

If photographs are to be used for publicity purposes, it helps if somewhere in the picture the company's name is visible on a sign board or a vehicle or item of plant.

There are many advertisement hoardings in the country which are in steady demand, frequently to advertise products. There is no reason for contractors to ignore the possibility of advertising on them. The cost is not great, but few people do more than glance at hoardings, so the most to be expected is that the company's name will register by being seen by a great number of people. Advertising on hoardings in general is not recommended, but special locations are worth consideration; railway termini, airports, seaports and similar places where many people gather and often have time of their hands and will possibly look more closely at such advertisements.

Movement in advertising signs is always helpful. It attracts people and they will stand and watch until the movement is complete. At the sort of place mentioned above, consideration should be given to using a machine such as a rotasign for advertising purposes. These machines which are electronically controlled, consist of a blind fitted on to rollers which move the blind forward and repeat the process over and over again. The blind can carry photographs in black and white or colour together with any descriptive material. They can show how a project progressed from start to finish or any other message of the contractor's choice. They give at least two year's service for an initial expenditure which depends on the message to be transmitted. The cost of the machine itself is reasonably cheap, and an annual maintenance contract can be agreed to ensure that the machine is regularly serviced.

In recent years advertising on football grounds, both association and rugby, and on county cricket grounds and race courses has developed considerably. The reason is the much greater television coverage given to these events. The procedure is for a board or screen to be erected inside the arena in a position where it can be seen by most of the spectators at the event, and where the television cameras will pick it up from time to time. Such advertising is not cheap, but the demand is substantial. It is a method of getting the company's name known, for the display boards cannot usually show anything more than the name. If the name is something like Watson Construction, so much the better because it describes the service provided.

There are some drawbacks to this sort of advertising. The choice of position is not always very great, and a board which is infrequently picked up by the cameras is clearly unsatisfactory. Advertising in this way on county cricket grounds is at the mercy of the weather. A whole match or part of a match can be washed out and the advertisement is then non-effective. In discussion with agents, this possibility should be gone into and arrangements made for a refund or other alternative recompense.

Transport undertakings such as the London Passenger Transport Board and similar organizations in other parts of the country offer opportunities for mobile advertisements. These take the form of a simple advertisement on the side of a bus or the complete advertising space on the bus. Buses can be painted in the company's colours and the whole external space then used to publicize the company. The cost of this sort of advertising is not exhorbitant, but care should be taken to ascertain the route which the bus will be taking. For example, will its route be confined to outer London, or will it regularly cross the central areas?

Other advertising opportunities arise from sponsorship which will be dealt with in chapter 6.

Advertising agencies

As manufacturers and distributors increasingly found it necessary to advertise their products or services, it was appreciated that this was a specialist field requiring specialist treatment. Consequently groups were formed of brokers who would act as intermediaries between clients and newspapers and other publications. This led to the establishment of advertising agents who offered their services to clients in the form of copywriting and graphic art services. As with most establishments with a common purpose and common problems, an association was formed with the current title of the Institute of Practitioners in Advertising. Most companies use advertising agents rather than attempt to write their own copy and place the advertisements themselves. This is obviously sensible because the need for the agency's services is not a continuous day-by-day requirement. A good agency will advise on choice of media and timing and will prepare copy for consideration and agreement. It will deal with the media direct and settle the account and subsequently invoice the client. This progression is illustrated in the box.

Information about space availability from publisher, i.e. newspapers, journals, year-books, etc., obtained by marketing director who informs advertising manager.
|
Consult advertising agent.
If considered a suitable occasion, discuss copy and cost.
|
Advertising agent produces draft copy for Advertising Manager/Marketing Director.
|
Advertising agent produces artwork of final copy for approval.
|
Advertising agent sends copy to publisher and invoice to contractor.

Arrangements made with the media by the agencies give them their main source of income. The agency is given a discount of from 10 to 15% by the media owners on advertisements placed, but charges the client the full cost. In addition, the client is also charged for the cost of preparing copy.

The usual way in which contact with an agency is maintained is frequent meetings between the marketing director of the company and a director of the agency who is responsible for the company's account.

If an agency is to be used for the first time, it is a good idea to approach the Institute for a list of suitable agencies and to visit their offices for discussion and to see examples of their work. Institute members will not accept accounts from competitors in the same industry for obvious reasons. Once an agency has been appointed, it will be necessary for the marketing director to discuss with the agency the nature and objectives of the company and for the agency to draw up a plan of campaign to be discussed with the company.

Most advertising agencies these days are prepared to undertake much more than the prepartion of advertisement copy and the placing of the advertisement. They will produce brochures, posters, leaflets and undertake direct mail shots and unless the company has staff of its own who are skilled in this kind of thing, it is best to use the agency's expertise.

Chapter 6

Public relations

Public relations is not easy to define because it embraces so many aspects of a company's activities. Its purpose is to create and maintain a climate favourable to the company. It is concerned with the creation of a good public image, and this requires a determined and continuous effort in correctly communicating to the public at large and to special sections of the public the nature, purpose, objectives and achievements of the company.

For whom is it necessary to create a favourable image of the company? Clearly this includes customers and potential customers. It follows that for contractors it includes architects, engineers and quantity surveyors whose advice to clients is important to the contractor. It includes also Ministers, Members of Parliament, councillors, shareholders and potential shareholders, subcontractors and suppliers, trade unions, workers already in the industry and students considering possible employment in the industry, its own employees and the general public. If all these people can be persuaded that the company is efficient in its management, successful in its operation, aware of its public responsibilities and has a social conscience, then its standing in the industry and outside it will be high.

There are many ways towards these objectives, all of which can be defined in one word – communications. Many of the troubles of the world result from poor communications. Statements not clearly made and consequently not clearly understood, messages or instructions capable of misunderstanding, failure to ascertain the facts of any situation, and general inaccuracy in communication can, and frequently does, lead to serious difficulties and even complete disasters.

In the following pages some aspect of public relations will be dealt with in some detail. These include the need for good communications and the methods by which these can be achieved. Much of the communication will be with the media – the press, television and radio, journals – members of the public, customers and their advisers and these communications will be by written or spoken word. It is important, therefore, to take the utmost care whichever method is used.

Communication within the company

For efficient operation it is essential that one's own employees should have no doubts as to what is required of them. This means that written or verbal instructions to employees should be clear. They must be clear to the instigator and to the recipient. The instigator has a duty to check that the message he wishes to convey is accurate and incapable of misinterpretation, and it is worthwhile checking that the recipient really does understand. Equally, it is up to the recipient to check with the instigator if he has doubts as to what is required of him.

External communications by telephone

Frequently the first contact made with a company is by telephone. The impression given to the caller depends on how his call is handled. Nothing infuriates callers more than to have to wait a seemingly long time before anyone answers. On receiving the call the operator should give the name of the company and the location, e.g. James Smith & Son, Chorley. This ensures, or should ensure, that the caller has the right number. He should then state his requirement and the operator should say that he is being connected to the person to whom he wishes to speak or who is dealing with the particular matter. If, after a reasonable time, the extension does not reply, the caller should be told and informed that he will be connected to the alternative extension. In all cases there should be an alternative extension, an essential arrangement in an efficient organization. If that, too, fails to reply, the caller should be informed and asked if he wishes to leave a message. If so, it should be written down by the operator, checked with the caller, and passed to the member of the staff or his secretary at once, so that it can be dealt with promptly. Slight variations to this procedure may be adopted, but the important thing is to organize the telephone system internally so that callers get the impression of an efficient organization. Inefficiency in the use of the telephone can lead to lost opportunities; efficiency can lead to increased business.

A useful, almost essential part of the communication system in an office is a public address service. This enables telephone operators to broadcast messages and particularly to locate staff out of their own offices when they are required on the telephone or for other reasons.

Although the telephone arrangements within a company are part of public relations, they are by no means the only aspect of this important subject.

Public relations are a means of promoting a company. They provide opportunities for informing the public of the purpose and achievements of a company in much the same way that advertising does, but without the space costs which advertising entails. Public relations cover a broad area. It includes any aspect of a company that impinges or could impinge on the general public, and particularly on prospective clients and their consultants. It starts with the telephone and goes through letter headings, envelopes, brochures and leaflets, press handouts, relations with the press, television, radio, contract progress, films, exhibitions, house magazines, annual reports, queries and complaints. In fact, it covers all types of communications about the policy, resources and activities of the company.

Some companies enjoy a public relations officer, not always responsible to the marketing director. Since public relations is just one aspect of marketing, it is best if it is clearly the responsibility of the marketing director, although a member of his staff may carry out the executive work under his direction. If there is a marketing director and an independent publications officer, there is a very serious risk that lines will be crossed and embarrassment caused.

Some companies employ public relations consultants of whom there are a great many. Most of them are very competent and experienced, and, because they have many clients, their contacts with the media are of great value. They can advise on the most suitable outlets, on timing and on the format of such things as press handouts. They also undertake the arrangement of press conferences, presentations, participation in exhibitions, site visits, opening or topping out ceremonies, the production of house journals, in fact any task which comes under the heading of public relations.

Although these consultants are experts in their own field, they are unlikely to be experts in the construction field. If they are employed they will, for some considerable time, have to be spoon-fed to ensure accuracy. This means frequent meetings between the company's public relations officer or marketing director and the consultant's representative handling the account. Unless the construction company is very large, and not many are, it is doubtful whether it is necessary to work through a consultant. Provided the company's own public relations officer makes it his business to meet the appropriate news, financial and industrial editors of the media, so that he deals with the appropriate person, there is much to be said for dispensing with the use of consultants. Much will depend on the frequency with which occasions arise when a consultant might be useful, but for most companies in the industry opportunities will not arise often enough to justify the not inconsiderable costs involved. There may well be special occasions when the employment of a consultant is justifiable, and it is a matter of judgement as to when outside help should be sought. Some aspects of public relations are dealt with in detail below.

Press notices

Possibly the bulk of the work which will fall to the public relations officer through the marketing director will be the production and circulation of press notices. These will enable the company to disseminate factual information about the company's activities, to give indications of changes of policy, and to correct any misunderstandings about the nature and purpose of the company which may be considered to exist.

It must be remembered that there are a great number of companies of all kinds, public authorities, and a host of miscellaneous organizations all using press notices to further their causes. Space in newspapers and time on television and the radio are limited and the editorial staffs of these establishments have the difficult task of deciding what to include. They have far more material than they require and so press notices are often consigned to the wastepaper basket, or at least condensed to only a very few lines. These important facts must be clearly understood by those who draft press notices. They will help in deciding the way the draft is prepared, what to include and what to omit.

A standard form for press notices is regarded as essential. It must show clearly the name of the company and its logo if it has one. It must state clearly that it is a press notice. It must show clearly when the information it contains can be published. It must show clearly to whom questions arising from it can be addressed with the appropriate telephone number. It must indicate how many pages there are in case a page or pages become detached. These basic requirements apply to all press notices whatever their subject matter. A typical press notice is illustrated in the box.

RAP	R. A. PARTRIDGE PLC 17–23 COMLEY ROAD ALICOMBE ESSEX 6EX 4RA	PRESS NOTICE No. 37 8 APRIL 1984

EMBARGO: NOT TO BE PUBLISHED BEFORE
0001 HRS ON 12 APRIL 1984

HEADING IN CAPITALS

TEXT

ENQUIRIES TO:
J. D. SMITH
TEL: 037-484 616

PAGE 1 OF 3

Because of the large number of press notices received every day by the media, if a company's handout is to have any chance of being used it is essential that it should be addressed to the correct individual in the media. It is, therefore, crucial to find out who is the appropriate person and to try to arrange to meet him or her. Personal contact of this kind is most important, and provides an opportunity to discover his likes and dislikes on the format of press notices and, more important, what day of the week he likes to receive them and the likely publication days. This latter is important because some newspapers have a special feature about construction and the aim should be to submit press notices with a view to inclusion in those features. In the case of weekly newspapers and trade journals, it is necessary to find out the last day for notices to be received for inclusion in the next issue.

Press notices should be as brief as possible but not to the extent of omitting important and more particularly interesting information. The writer should try to envisage the sort of copy which readers would find interesting, bearing in mind that although only about one in ten of the working population are regularly engaged in the construction industry in its widest sense, there are many more who pay rates and taxes and are broadly interested in the way the environment is being developed. Since editors and their staffs are busy people with very many notices to read, it is as well to try to get the main story into the first paragraph, which has the best chance of being read. Furthermore, a striking headline will often encourage further reading, and although the media seems to pay more attention to failures and disasters than to successes – the collapse of a building is more likely to be considered newsworthy than the successful completion of a new one – they do not ignore the latter.

What subjects can best form the basis of press notices? There are many and the marketing director/PRO must determine what should be dealt with in this way and when, as well as the extent of the distribution. In fact, the distribution need not be the same for all notices, but it is best to remember that there are many people and organizations interested in the company's activities: shareholders, investors, institutions, finance houses, clients, competitors, sub-contractors, suppliers, consultants, creditors, Members of Parliament, town councillors, employees, prospective employees, students, land and property agents, sales staff in other industries – the list is endless. It is important, therefore, to compile a list of newspapers and journals which might be included in a distribution list to be used on most occasions, even though it may not be followed blindly on every occasion. There may be times when a shortened version of the distribution list seems adequate and at other times some aditions to the list might seem sensible. One thing is certain, if a newspaper or journal does not receive the notice, they will not publish it, so over distribution is better than the reverse.

Some of the subjects which might be covered by press notices are discussed below.

Contract awards

A good indicator of the progress of a company so far as work load is concerned, is the number, type and size of contracts awarded. It is important that the marketing director is informed immediately of the details of contracts awarded to the company and its subsidiaries and its associated companies if there are any. He can then decide on the need for and the nature of a press notice and on its timing. The size of contracts worthy of a press notice will depend on the size of the company. There is no point normally in a company with an annual turnover of several millions of pounds issuing a press notice to announce the award of a contract for £20 000. For a much smaller company operating in a restricted local area, this may be worth publicizing. Even for a large company it may be worthwhile issuing a press notice for such a contract if there has been particular local interest in the project or if it has been the subject of local or even national controversy. A judgement has to be made as to its newsworthiness, but contracts of £250 000 are regularly reported in the press and technical journals and should be considered for inclusion in press notices.

There are two schools of thought about press notices announcing the award of contracts. One school believes it wise to wait up to a month so that the total of contracts awarded is sizeable and then issue a notice announcing the award of contracts totalling several million pounds, but listing in the notice those contracts of, say, £250 000 and over separately. The other school believes it better to issue a separate notice for each contract of, say, £250 000 or over as soon as it is awarded. This school argues that several announcements are published by this method as against one by the alternative. Again, decisions will be affected by the size of the company and the nature of the individual contracts. One the whole the first method is favoured generally because the award of contracts amounting to millions of pounds is more striking and will make a greater impact than the award of smaller contracts noticed individually. Individual marketing directors/PROs should make their own minds up about the method to be used and should judge the effectiveness. It is useful to arrange with a press clipping service or a PR consultant to send to the marketing director clippings of the publicity obtained by the issue of press notices and in the light of these the distribution list may need revision from time to time.

A typical press notice announcing the award of contracts is given on page 70. The text should be drafted in a form likely to be of interest to all on the distribution list, but some newspapers which run special features one day per week on construction and most technical journals are more interested in detail than others. This problem can be dealt with by using the same press notice for all on the distribution list with an appendix of additional information for those who like to have it. To some extent this applies also to local newspapers where more local information is considered newsworthy. Local radio stations are interested particularly in personalities and are more inclined to use press notices that name local people who will be connected with the project.

It is always advisable to give publicity in these sorts of press notices to clients and particularly to consultants. Until recently consultants were severely restricted in the amount of publicity they could engender for themselves and even now contractors will be doing themselves no harm in publicizing the consultants in this way. They may be doing themselves quite a lot of good.

Care must be taken to respect the wishes of the client about publicity. Most clients and consultants have no objection to their contracts being included in contractors' press notices. There are, however, occasions when the client has requested no publicity and this must be respected. This request is sometimes stated in the bill of quantities, and sometimes it is a verbal request. In the latter case whoever in the contractor's organization receives this request must pass it on to the marketing director/PRO at once and without fail. Sometimes the client has no objections to a vague description of the project being included, but has strong objections to any detailed information about the purpose of the project. For example, 'a factory at A' is permissible, but 'a factory at A to manufacture a named product' is not. Some activities or proposed activities are not to be released to competitors or others, and this is understandable.

There are occasions when a single contract of national importance or interest warrants a press notice on its own. Examples spring readily to mind: a new section of motorway, an important new bridge, a major new factory promising high employment opportunities. In such cases, the client may wish to issue a press notice to publicize the award of the contract. It would not be sensible in these circumstances for the contractor to issue a separate notice and it is better to co-operate with the client and to try to ensure that the client's notice includes sufficient detail for the technical press and that they are included in the distribution list. Clients expect approaches of this kind and tactfully handled, no particular problems arise. The example in the box is of a typical

£8 m of new contracts for R. A. PARTRIDGE PLC

R. A. PARTRIDGE PLC the national building and civil engineering contractors of Alicombe, Essex, announce the recent award of contracts totalling some £8 m for public authorities, industry and commerce.

Included in this total are the following contracts:

(1) Bypass at Selfod for Wanton County Council
Value £4.2 m
County Engineer: W. S. Johnson B.Sc., F.I.C.E., F.I.H.E.
Work will start on 23 April and the contract period is 18 months.

(2) Factory at Wineton for Challenon Soaps PLC
Value £1.8 m
Architect: Wilson & Associates
Structural Engineers: J. D. Brown & Sons
Quantity Surveyors: W. A. Pastow & Sons
Work has just started and the contract period is 9 months.

(3) Shops and offices at Tickworth for Brumskill Developments Limited
Value £1.4 m
Architect: Designers & Planners Associates
Structural Engineer: H. C. Wrigley & Sons
Quantity Surveyors: J. K. Jerome & Associates
Work will start on 1 May 1983 and the contract period is 18 months.
This contract was awarded to Partridges South West Region located at Westonport.

Enquiries to: J. D. Smith. Tel: 037-484 616

Additional information

(1) The bypass is nearly 1 mile long and extends from Brian's Bridge to Crompton Cross. It will be a two-lane dual carriageway with no intersecting roads, and constructed with flexible pavement. There will be pedestrian footbridges with ramp access at Burrows Road and Digneys Road. The Agent for Partridge will be Peter Morton B.Sc., M.I.C.E. who lives at Selford.

(2) This factory will be steel framed with a flat roof in steel and with profiled steel cladding above a 2 metre brick wall. It will measure 200 m × 36 m and 10 m high and incorporate a 36 m × 40 m offices and amenity block. External works include drainage, service roads and concrete finish parking for 80 vehicles.

(3) This development is in Wordsworth Road, Tickworth. It will include 12 shops with offices on two floors above. Construction will be of reinforced concrete frame with brick cladding. The scheme includes parking facilities in front of the shops for 20 cars and there is to be a car park for 80 cars at the rear of the premises.

press notice about the award of contracts. All the places and names in the specimen are, of course, entirely fictitious, and have been included to show how such notices can be compiled.

Contract progress

Contract progress can usefully be divided into four parts:

(1) Commencement on site;
(2) Completion of shell;
(3) Handing over of project;
(4) Official opening.

Each of these stages provides an opportunity for publicity and if this is undertaken it will usually mean a ceremony of some kind. Sometimes the client will feel that he wants publicity for the project for the purpose of indicating the contribution the completion of the project will make to the lifestyle of the general public. Frequently public authorities, central and local government, development corporations and nationalized industries, want publicity to demonstrate the progress they are making towards some desirable objective. Industry and commerce sometimes wish to inform the public of some new facilities which the completed project will provide. Sometimes a project creates a great deal of general interest and the public likes to be kept informed of progress on such projects. Whatever the motive power which prompts such wishes, the contractor's work and company will obtain some free publicity. Free is used in the sense that no payment is involved in distributing press notices, unlike advertising. Nevertheless, some costs are involved in preparation and the contractor frequently becomes financially involved in providing mementoes or refreshments.

It is, of course, open to the contractor to be the prime mover in organizing such ceremonies, and there are endless opportunities for doing so. Whoever sets the ball rolling in these matters, whether it be the client or the contractor, the prime need is for close collaboration between the two.

The procedure is of a standard form. The usual practice is to invite some person of prominence to perform the ceremony or to take a leading part in it. In some cases it may be important enough to invite a member of the Government and this means arrangements must be started well ahead of the desired date. Approaches to members of the Government should be formal and dates that are convenient for Ministers must be accepted. Sometimes the particular ceremony will have to be fitted in with some other engagements in the vicinity. It is usually the client who will make approaches to Ministers, but there is no reason for the contractor not to do so if he wishes. It would have to be an important project before such an approach was even considered.

Much more frequently it is the chairman or managing director of the client who is invited as the principal guest in the case of industrial or commercial projects. For publicly financed projects the chairman of the authority, the appropriate Minister, a local Member of Parliament or some other prominent dignitary should be invited. The value of inviting the local mayor or chairperson of the district council should not be underestimated. The local newspapers keep a very close watch on forthcoming engagements and his or her activities receive a very good coverage in the local press and on television and radio.

The ceremonies themselves are not very difficult to arrange, but protocol must be closely observed. Offence can easily be caused if someone who ought to have been invited has inadvertantly been overlooked. Liaison with the client on such matters is of vital importance. Even inside the contractor's company feelings can run high if the wrong choice of representatives is made. There is a tendency because the work is under way to think that so far as the company is concerned the only ones who need to be invited are the chairman, managing director, production director and site staff, and not only senior site staff. There is a tendency to overlook those who were instrumental in obtaining the opportunity to quote for the contract, those who prepared the successful tender, those who purchased the materials, and those who will have dealt with the financial side of the work. This undue emphasis on the construction side is thoroughly bad and resented by those who have made the initial contribution to the award of the contract or will contribute to its successful completion although not part of the site staff.

Once the skeleton of the ceremony has been agreed, the detailed work can proceed. The precise form of the ceremony should be settled and its location determined. Any equipment needed should be acquired and arrangements made for it to be available in good time at the right place.

Invitations should be sent out in good time and replies listed. If mementoes are to be handed out, these should be obtained. If there are to be refreshments or a luncheon, the necessary purchases or reservations should be put in hand in good time.

It is advisable to draw up a programme and a check list of things to be done. All this must be under control of the marketing director or the PRO. When the arrangements are reasonably finalized (they never are finalized until the last moment, usually because someone cannot come) the press notice draft can be revised to its final form. As soon as the date and time of the ceremony have been settled, it is advisable to inform the press and technical journals so that they can make arrangements to cover the occasion. They should be told that a press notice will be issued in advance of the date of the ceremony and when their copies are sent they should be accompanied by an invitation to the ceremony and to any refreshments. Obviously not all the press so invited will attend, but some will and they are much more likely to respond to an invitation than to the mere receipt of a press notice.

The text of the press handout is not very difficult to draft. It should have a heading which will arouse interest, e.g. 'Work to start on factory to employ 200', or, 'New hospital to serve West Lincolnshire to be completed in three years' time', or 'Work starts on new science building at 'X' university'.

The first paragraph must contain details of:

(1) Who will perform the ceremony;
(2) Where;
(3) When; and
(4) Brief description of purpose of project, approximate cost and name of contractor.

The rest of the notice should be devoted to any background information about how the project came into being, who the consultants are, etc. This can be followed by a brief description of the project, and the names of the consultants' supervising officers and the senior site managers for the contractor.

Topping out, completion of construction, handover, official openings

Topping out, completion, handover and official opening ceremonies can be dealt with in a similar way. As general guidance, consider these points:

(1) Has the correct chairman of the District Council been invited? Boundaries need to be considered. Have his initials or forenames been correctly stated, and his honours or qualifications?
(2) Should any officials of the council be invited and, if so, whom?
(3) Should the chairman or other representative of the county council be invited? Were they the planning authority or was planning delegated to the district council?
(4) Are ladies to be invited? If so, how many? What about toilet facilities on site?
(5) Are photographs to be taken? If so, solely by the press or the company's photographer?

> (6) Will police assistance be needed for traffic control? Should a police representative be invited?
>
> (7) Where will cars be parked? This should be as near to the place of the ceremony as possible.
>
> (8) When guests arrive, will there be a building or a marquee where they can have coffee or tea?
>
> (9) If it is a road or motorway project, is it necessary to have a coach to convey guests on a tour of inspection?
>
> (10) Is a tape to be cut? If so, are special scissors necessary?
>
> (11) If a plaque is to be unveiled, is there to be a curtain to be drawn? Does it work?
>
> (12) If concrete or bricks are to be placed, or earth turned, have the necessary tools been obtained?
>
> (13) If a memento is to be given to the person performing the ceremony, what form will it take and does it need inscribing or does a card need to be provided?

There may well be other matters to consider, but *all* the arrangements must be left in the control of the PRO working to the marketing director. If several people are involved, lines will almost certainly be crossed.

The box below contains an example of a press notice about an official opening. All names and places are entirely fictitious, of course, and this example obviously is not the only way to write press notices. It is given only as guidance, perhaps for those who have not so far been involved with such notices.

> **First supermarket in Highley opened**
>
> Councillor James H. Dutton, chairman of Wilton District Council officially opened the Sellwell Supermarket in Rook Street, Highley at 11.30 a.m on 12 November 1982 in the presence of Mr Thomas Crabbe, chairman of Sellwell PLC and a large crowd of local dignitaries and residents.
>
> This is the first supermarket to be opened in Highley, and will be a great convenience for local residents who previously have had to travel five miles to the nearest supermarket at Exley.
>
> Councillor Dutton offered his congratulations to Sellwell for their enterprise and praised the contractors, B. H. Osborne, PLC, of Highley, who had completed this £1m project in only 11 months. The store will save local residents journeys to Exley and will provide in addition to food a variety of household requirements, many of which were not readily obtainable locally.

Work started on this project in January to the design of Watson & Greene, the well known London architects. Mr George H. Watson, F.R.I.B.A., senior partner, was the controlling architect assisted by Mr Henry Wilson, F.R.I.B.A. The structural engineers were Thompson & Sons of Exley and the Quantity Surveyors, Shrimpton & Partners, also of Exley.

The store is a two-storey building. The ground floor, which measures 60 m × 40 m, will provide the shopping area for foodstuffs. The upper floor comprises the shopping area for household goods and the offices and amenities including a self-service restaurant. Storage space is provided at the back of the premises and there is concrete hard-standing for ten large goods vehicles and 150 cars. Loading bays will give easy access for deliveries.

The building has a reinforced concrete frame and is clad in brickwork with a flat roof. Shoppers can enter the building from Rook Street or from the car park. The location of the store is not likely to cause any delays to traffic passing through Highley.

Mr Basil Osborne, Chairman of B. H. Osborne, PLC, the main contractors, said he was delighted that his company had been able to complete the building ahead of time in spite of some very wet weather in the early part of the year. He congratulated Jim Brady, the agent, and his site team on a good contract successfully carried out.

Half yearly and annual reports

Opportunities for publicity occur when half yearly or annual reports are ready for issue to shareholders. Obviously, shareholders, the financial editors of newspapers and journals, the institutions and the general public are most interested in the details of performance in the period under review, and these details are included in the company's reports in detail. What is of special interest is whether there has been an improvement in performance or not and this is measured in some way or another. The best indicator is net profit as a percentage of capital employed, but this is not the whole story. Another important indicator is whether turnover in real terms, i.e. after allowing for inflation, has increased or not. In the case of a public company, asset value per share is also an indicator of the company's well being. It is, however, not part of the purpose of this book to deal in detail with the strictly financial performance of a company: that is better left to the financial director. It is the purpose to try to show how publicity for the company can be created when these reports are available.

Firstly, the form of the annual report itself is important. A well designed annual report creates a good impression even if the statistics it contains are not particularly good. So it is well worthwhile giving a good deal of thought to the format of the report.

The report should have a distinctive cover, preferably in the company's colours; by this means it is instantly recognized. It should clearly state on the front cover that it is the Annual Report with the appropriate year for the company. Ideally, it should say nothing more, but some companies include a colour photograph of one of their projects.

In addition to the statistics which form the bulk of the text of the report, it should contain details of the Annual General Meeting and the Director's Report.

The Director's Report should follow the details of the AGM closely and not be preceded by the tables of statistics. What readers of the report want to see first is the view of the directors on performance over the period, and particularly the prospects for the future. It is important that the marketing director should be given the task of drafting the director's report for consideration by the full board. The marketing director will know what aspects of past performance should be highlighted and which should be treated in a lower key. With regard to future prospects, he should be best equipped to comment, and a wise chairman will pay much more attention to the marketing director's views than to the views of other directors who are experts in some other field of activity.

All reports are dull to some extent, but they can be livened up by the inclusion of suitable colour photographs of projects completed or under construction. It must be remembered that most public authorities and some private consultants require copies of annual reports when determining lists of approved contractors. They need them to examine the financial soundness of the company and will be most interested in the statistics. Their judgement of a company's suitability for inclusion in their lists will be influenced, perhaps a great deal, by well chosen photographs of successful projects completed or others under construction.

It is useful when half-year or annual performance figures are available to take space in the press to announce to readers how the company has fared. Of course, the results will be included in a press handout, but newspapers, already blessed with a surfeit of similar information about many other companies, will normally report only the figures and a brief comment. To obtain wider publicity it is best to pay for an insertion and this should contain the important figures in the reports and a precis of the director's report or at least the salient points from it. The advertisement need not comprise anything more than those matters, but there is nothing to prevent the inclusion of a few black-and-white photographs of projects completed or under construction. The important thing is that any such advertisement should be left to the marketing director to draw up in conjunction with the advertising agent.

Press conferences

Occasionally, and it is only occasionally, there may be an item of such importance that it warrants a press conference. The acquisition of companies, mergers, substanial re-organization, the award of a contract of great importance or a serious disaster may be of such interest to the press and may be likely to give rise to numerous questions if dealt with solely by means of a press notice. In such cases, it may be better to hold a press conference to which industrial editors are invited at an appropriate time. Some modest refreshments and satisfactory seating should be provided. Which members of the press should be invited would depend on the importance of the matter, whether it were of national or only local interest. At these conferences, the chairman or the marketing director should make a statement of the facts and then invite questions. Invitations to the press to attend should, of course, be accompanied by a brief note of the purpose of the meeting and sufficient information to enable them to decide the sort of questions which they may wish to ask.

Similarly, it may be useful to have a press visit to some important project on its completion or at some appropriate early stage. If some unusual feature in the construction of the project has been employed, or if there is great interest in the project, the press like to have an opportunity of such a visit. The same idea applies to visits by professional organizations whose programme secretaries are keen to arrange appropriate visits for their members. Requests for such visits should normally be accepted with alacrity and the company should extend some hospitality to the visitors and arrange for suitable promotional literature to be available.

Presentations

Some companies invite clients, architects, engineers and quantity surveyors to gatherings at which the virtues of the company can be extolled. These are useful platforms for introducing new directors or managers to the professions and clients, but the real purpose is to promote the company. Arrangements are made to hold these gatherings at lunchtime and to provide refreshments. It helps to secure a good attendance if a well-known speaker, not necessarily involved in construction, is invited to speak on some matter of general interest. For example, for such a gathering to be addressed by a top flight golfer, cricketer, yachtsman, author or explorer would probably increase the attendance considerably and the result would be to the benefit of the contractor. It is important that such gatherings should be run to a close timetable so that those who attend can meet their afternoon engagements without having to leave before the end of the gathering.

Presentations of this kind are most useful in areas in which the company is not particularly well-known and where it is hoping to expand its business.

It is vital to keep a record of those who attended and to maintain contact with them. A letter of thanks for attending, accompanied by some company literature, is well worthwhile.

Invitations to sporting events

Now that many football and cricket clubs have built executive boxes on the grounds, and the Rugby Football Union have followed the Welsh by issuing debentures entitling the purchaser to reserved seats at Twickenhan for international matches, there are many opportunities available for the entertainment of friends and prospective clients. It is now commonplace for contractors to take advantage of these facilities. It would be wrong to assume that friends so invited would be unduly influenced by such invitations in their workaday life. These arrangements simply serve to cement friendships which are so helpful in the successful carrying out of contracts. Contracts are still awarded on a competitive basis and no contractor using these methods of creating goodwill should expect any favours in consequence, but they do help to form friendships, some of which last for a lifetime.

Sponsorship

From time to time contractors will be approached to see if they are interested in sponsoring some event or some activity. This may be no more than lending one's name to a particular endeavour with some financial help such as donating a couple of prizes at a local horticultural show. On the other hand, it may be much bigger and involve a considerable outlay. For example, major cricket and other sports competitions are sponsored by substantial companies at very great cost. Some companies are heavily involved in sponsoring horse racing. The objective is undoubtedly to help the particular endeavour, but at the same time it is a significant form of advertising.

Such activities do not stop at lending one's name to the endeavour and contributing to the cost. They involve entertainment in many cases and provide opportunities of meeting people of influence which may benefit the company in the long run. The marketing director will be principally involved in such arrangements. He will have to decide how best to advise his managing director or chairman on such approaches and, in doing so, he will have to give an estimate of the total cost. In assessing this, he should not overlook the entertainment cost because this can be very substantial.

Brochures and sales literature

Some companies employ representatives who travel around paying visits to clients and their consultants to seek opportunities for contracts. Others seek business differently and the relative merits will be discussed later. Whatever method is used, there is no doubt whatever that sales literature in one form or another is essential.

The most obvious form of literature is the company brochure, a comprehensive catalogue of the company's history, resources and achievements. It goes without saying that the compilation of such a document is a matter for the marketing director.

What form should it take? Obviously it should be a booklet in A4 size. This is the usual size of most magazines and journals and is by far the best size for a brochure. It should have a distinctive cover in the company's colour, the front cover bearing the company's name and possibly the address and telephone and telex numbers, although these may be included elsewhere in the brochure. The inside of the brochure should contain:

(1) The names of the directors.
(2) A brief history of the company, highlighting some of its more outstanding achievements.
(3) An up-to-date statement of annual turnover and number of staff and operatives.

(4) Addresses of outstations with the names of local managers.
(5) Types of work undertaken and where, with a note about any specialized work.
(6) A note about industrial relations.
(7) A note about safety, health and welfare.
(8) An organization chart.
(9) A typical site organization.
(10) A list of clients for whom contracts have been undertaken.
(11) A list of consultants worked for.

This information can be given in two or three pages and should not be allowed to extend to more than this.

The rest of the brochure should consist of photographs of projects successfully completed or under construction. These should be grouped under various headings, such as offices, education, factories and storage, commerce, motorways, roads and bridges, sewage disposal, hospitals, leisure, and any other appropriate groupings.

The choice of photographs should, in the main, not include projects completed more than, say, five years earlier. Prospective clients are much more likely to be anxious to see what the company has achieved in the recent past, rather than in the distant past. Each photograph should be properly captioned with its name, the client, the names of the consulants, the date completed, the approximate cost at the time and the contract period.

Photographs of projects of particular interest might have a note describing in more detail what the interest is.

It is worthwhile including in the brochure the names and functions of subsidiary or associated companies, but photographs of their projects should be left for inclusion in their brochures.

There is much to be said for brochures being bound in a loose-leaf style. With this type of binding pages can be replaced whenever necessary, and this applies particularly to photographs.

The whole brochure should not be more than 10–12 pages in length. Anything bigger is largely a waste of paper and if sent by post becomes costly to deliver.

Sufficient copies should be printed to last two years and then a revision should be undertaken. Up-to-date brochures are more likely to show the real position of the company and brochures several years old create a bad impression.

Some companies produce colour photographs of projects as soon after completion as possible and have these printed on cards together with photographs taken at the start of the contract and at appropriate stages of construction showing the actual dates. The photographs are accompanied by text which is descriptive of the contract and how it progressed to completion. These are sent out to people on a standard mailing list determined by the type of project. For example, factories are sent to industrialists and their clients, hospitals to health authorities and so on.

Other forms of trade literature may become necessary from time to time. For example, if it has been decided to try to get contracts on a design and construct basis, or on a prime cost plus fee basis, it is advisable to produce a short brochure to advise clients that these services are available. Naturally they will be shorter than the main company brochure and may be no more than a four-page folder. They should, of course, set out clearly the service being offered and if examples can be given of contracts successfully carried out by these methods, they should be described and illustrated by plans and photographs. The essential thing is to show not how the contractor benefited from these methods, but how the client benefited in time and/or cost.

Films

There are many films available about various aspects of construction. These can be hired and are very useful, particularly for training purposes. Films can also be useful for promotional purposes, especially if they show the way an important project was constructed, and especially if the construction incorporated some special technique seldom or never used before in the United Kingdom.

The bills of quantities for some important projects include a requirement to produce a film showing the actual construction progress. These films are a client's requirement, but clearly can be used as promotional material by the contractor.

Opportunities for filming construction work are endless, but not all projects are suitable for this treatment. Contractors should bear in mind the possible advantages to their company which could arise from a well-produced colour film with suitable commentary. These films can be shot by the company's own photographer, or by a film company who specialize in this field. If outside help is used, it is important that the company should appoint someone to accompany the photographer each time he shoots some footage. The best person to do this may be the agent in charge of the project, or whoever is in charge of public relations. It is important to ensure that the best features of the construction process are shot and from the best viewpoints.

More film is likely to be available than is required and editing is important. A selection of film to be used can best be made by the public relations officer/marketing director and the production director. It will also be necessary to compile a commentary and to put the work of producing the final copy out to one of the specialist firms who will arrange for a suitable person to be employed to give the commentary.

This promotional use of films will not arise in many companies for lack of projects of sufficient importance or interest. If the opportunities do arise, they should be given serious consideration because a well produced film can be used extensively for promotional purposes and enables the company to organize film shows to which prospective clients and their consultants can be invited. The films can also be used for recruitment and training.

The cost of producing a film is not excessive bearing in mind the amount of publicity it can engender. The cost compares quite favourably with the cost of advertising, and, of course, the audience is hand-picked.

Exhibitions

There are many exhibitions concerned with construction held in the United Kingdom, notably the Interbuild Exhibitions at the National Exhibition Centre and in Scotland. Companies and organizations from every field of activity concerned with construction are represented at these exhibitions and no doubt many orders, or enquiries which lead to orders, are obtained, particularly by those exhibitors who produce or sell materials and equipment for construction.

Some contractors have stands at such exhibitions, but apart from keeping their names before the prospective clients and consultants who visit the shows, there is little they can do except to show photographs of their projects and have ample supplies of promotional literature available for visitors. It is very doubtful whether contractors who take stands really get value for money. The costs include space rental, stand equipment, staff costs and entertainment. These can be substantial and the return for this expenditure is difficult to assess. On the whole, it is not the best way for contractors to publicize themselves. There are better ways of promoting contracting companies, but perhaps companies should consider taking part, or even take part and reach their own conclusions.

House journals

Earlier in this chapter it was said that public relations concerned relations with the public generally and with one's own employees, and so it does. It is as important to keep one's employees informed of the company's policy, objectives and achievements as it is to inform the general public. The company is the bread and butter of its employees, and they must be kept informed of its progress. Indeed, it is a serious mistake to withhold from employees information that they can be given without risk, and keeping information confidential unnecessarily gives rise to rumours and wrong impressions which could be avoided. The policy best followed is one of openness, and wise directors seek the views of their senior staff on policy matters to be discussed at board meetings.

One way of keeping employees, and particularly staff, informed is through the medium of a house journal produced at intervals and giving concise information about policies, activities and people, especially about people. In large companies employees tend to be moved about from office to office and site to site. These employees are known to many members of the organization and their colleagues like to know where they are and what they are doing. The information need not be confined to the work of the company. There is room for news of colleagues' domestic and leisure activities. There is also room for listing new appointments, departures including retirements and deaths. There is also room for engagements, marriages and births, but it is company news which really justifies the production of a house journal.

In small companies, the house journal need be no more than a typewritten document of a few pages. In large firms it might carry photographs (quite well produced by photostat processes) and be bound inside a simple cover. This is not expensive and really involves little more than the staff cost of the editor and his secretary. Colleagues should be encouraged to submit information considered suitable for inclusion, and if there are outstations or associated companies they should be required to submit contributions. There is much to be said for not having regular publication dates, but to have issues only when sufficient material to warrant this is available. Otherwise publication dates arrive when there is little to say.

In large companies it may be advisable to employ an editor whose sole job is to produce the house journal. In such cases it may be advisable to produce issues at regular intervals, monthly, quarterly or twice in a year. It may also be worthwhile having the issue printed with photographs in which case outside help in production will almost certainly be necessary. Some companies circulate house journals to prospective clients, consultants and competitors. This increases costs, but it is another means of publicity. Some house journals include advertising space, the payments for which reduce the production cost.

Public relations consultants and some advertising agents are ready to undertake the production or circulation of house journals. They do this in a professional way and the cost is quite substantial. If this method is adopted, it will still be necessary for the contractor to appoint an editor to supply the basic material and to check the draft produced by the consultant. It is better if the whole production can be kept within the company, and the marketing director and his staff are capable of producing a house journal perfectly adequate for its purpose.

It is always useful for the company to attend annual dinners and other functions of the trade organization of which it is a member, and to take along guests from amongst clients and consultants. These are good opportunities for forming useful friendships. Similarly, the company should encourage its directors and staff to play a full part in the activities of the trade and professional organizations of which they are members. Useful information and contacts are made in these ways.

Logotypes

Logotypes, more commonly called logos, are symbols used by many companies and other organizations on their letterheads, and in their publicity material. They appear in a wide variety of forms and in recent years have been used by an increasing number of organizations. Their purpose is to associate the name of the organization with the logo, so that the sight of the logo immediately brings to mind the name of the organization.

The logo in its simplest form may be no more than the name of the organization, but if so it is always depicted in a distinctive way, by shape or colour. It may be merely an abbreviation of the name of the organization as, for example in:

It may be just a set of initials as, for example, in HMSO, PSA or

Some organizations adopt shields or coats of arms for their logos, whereas others prefer a symbol of some kind. There are a host of examples which are well known and instantly recognized. Examples are National Westminster Bank, Tarmac, Taylor Woodrow, Barratts Oak Tree, Bovis's humming bird, British Steel, and British Rail.

Companies in the construction industry might consider whether the use of a logo would be beneficial to them. If they decide to go ahead they should employ a graphic artist to produce possible designs and then choose the one they like best. A simple logo, easily recognized and unlike those of their competitors, should be chosen. Once the choice has been made the logo should be used on letterheads, brochures, in advertisements and in other promotional literature. The aim should be to so familiarize everyone with the logo that its appearance immediately brings to mind the company's name and function.

In this chapter some aspects of public relations have been dealt with in some detail; there is no claim that the subject has been dealt with in a comprehensive way. Matters will arise where careful handling of relations with the public are of the utmost importance. It is hoped that what has been said here will have been sufficient to emphasize that good public relations are vital for the progress of a company: bad public relations lead only to reverses.

Chapter 7

Securing opportunities for contracts

Perhaps the most important aspect of marketing is ensuring that sufficient opportunities are obtained for the award of contracts which will enable the marketing plan to be achieved. Certainly this aspect of marketing is regarded by many contractors as the sole function of marketing staff. They have not yet appreciated that forward planning is an important management function, nor have they accepted that avertising and public relations are vital features of the marketing concept.

The days when invitations to tender or unsolicited orders came to contractors fairly regularly without any effort on their part are gone. The shortage of work in the last few years has demonstrated this without any doubt, and contractors have to make determined efforts to obtain opportunities to quote; they have to sell their services and, like all other industries, they have to keep themselves informed as far in advance as possible of forthcoming needs for construction services.

In an earlier chapter the best sources of information about future projects has been dealt with at some length. Now it is necessary to say what action a contractor should take to try to ensure that he will be given the opportunity to quote. These are summarized in the box on page 87.

Some contractors, indeed many contractors, employ salesmen or representatives, often under the title of marketing manager, whose sole occupation is to find out what projects are being considered or are in various stages of planning. Their principal method is to make regular calls on architects, consulting engineers and quantity surveyors in private practice or in the public service

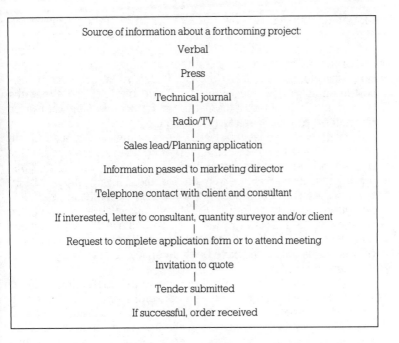

and to endeavour to ensure that their firm is given the opportunity to quote. In companies with regional or area organization, this task is allotted to the local marketing man or it is sometimes undertaken by a member of the estimating staff.

Consultants react unfavourably to visits by salesmen and there are several reasons for this. The obvious one is that the partners or the project consultants are busy men trying hard, often desperately hard, to meet deadlines laid down by their clients. Anything which interferes with their work is not welcomed. This often means that the salesman is kept waiting and, in the end, does not see the person with the power of decision. This happens even when a definite appointment has been made, and there is little the salesman can do about it.

Moreover, the salesman often cannot speak with the necessary authority. He cannot say with certainty that his firm will quote if given the chance. Much will depend on the load of work in the estimating branch and even the technical problems of the particular project. It is not unknown for an opportunity to quote being obtained only for it to be rejected for reasons which should have been clear before the approach was made. In some cases tender documents received are returned because of misunderstandings or failure to keep close enough to the marketing plan.

These visits by salesmen often result in reports by them of firm opportunities to quote for a number of projects spread over the next six months or even longer. They are indications by the salesmen as to how successful they have been. In the event, the opportunities, or some of them, are not forthcoming for one reason or another, or if they are forthcoming they have to be rejected by the firm for adequate reasons, or sometimes for inadequate reasons. The importance of the marketing director's role in matters of this kind must not be overlooked, and once he has decided that a particular project is one to quote for, the decision should not be reversed except for reasons of inadequacy of estimating staff. It is not often sensible to reject the opportunity for any production reasons. If the marketing plan is to be achieved, the production function is to carry out the work successfully and the adequacy of production facilities will have been taken into acount in drawing up the marketing plan.

The employment of salesmen, i.e. local marketing staff, is something which requires careful consideration and it is interesting and informative to note how the average salesman's time is spent. Whilst it is not claimed that the figures given in the box are absolutely accurate, they give a good indication of the proportion of his time actually spent in selling.

Travelling time	30%
Waiting time	15%
Parking time	5%
Breaks	15%
Selling time	25%
Reporting time	10%

If each interview takes on average half an hour, the number of customers seen each day is no more than three, and with traffic and parking problems as difficult as they sometimes are, there must be some doubt as to whether even three customers can be seen.

The cost of employing salesmen of the right experience and calibre is likely to be high. They need to be people with considerable experience in the industry, preferably professionally qualified. They must be mature and so no younger than about 35, and preferably a good deal older so already earning a reasonable salary. They must dress well and be good communicators already known personally to a good many clients and consultants who must respect and have confidence in them.

It is difficult to see how such a person can be obtained for less than £10 000 per annum and with on-costs the total cost to the company must be nearly twice that figure. He or she must be provided with a good car, either purchased or on fleet hire, and expenses will have to be paid.

The cost of employing this method of securing opportunities has to be compared with other methods and on the whole the direct mail approach is considered to be less costly and more successful, as will be shown below.

It has long been the practice of public authorities to keep approved lists of contractors and a company should endeavour to be included in these lists. This involves usually completing an application form, a process which must be done with great care. In recent years some clients and a good many consultants have adopted the same method either for the preparation of an approved list from which selections are made for tender lists as projects reach that stage, or for each individual project at the appropriate time. Even companies of international repute and very well-known local companies are not excluded from these procedures. It is not possible to deal with these applications by a standard reply as the questionnaires differ quite considerably from organization to organization. Each must be dealt with individually, and care must be taken not to miss the reviews which some public organizations undertake every three years or so.

In some cases, admission to the approved list will mean an unsolicited enquiry in turn. Some authorities operate a system of including in a tender list the two lowest tenderers for the last similar contract, with the remainder of the tender list being selected by rotation from firms on the approved list considered capable of undertaking the work. In spite of this it is always advisable to speak on the telephone to the authority or consultant expressing a wish to be included in the tender list for a particular project. Telephone operators in the consultant's office are very helpful in providing the name of the partner or project manager of the job in question, and it is usually not difficult to find an opportunity to speak to him personally. By so doing, further information about the project, its size, method of construction and planned tender and starting dates can be obtained which help in deciding whether to try to get on the tender list. These telephone conversations almost invariably terminate with a request from the consultant to send a letter indicating the company's wish to be given an opportunity to quote. Not all such letters are successful in securing the company's inclusion in the tender list, particularly as in some cases as many as a hundred firms have similarly applied. Much depends on the letter sent, and there is a good deal of skill and research involved in composing such letters. In a day it is possible to deal with at least eight such projects, or a weekly total of 40. The cost per project is partly salary and on-costs, telephone, secretary's typing and postage, together with any literature included with the letter. In other words, it is very much less than employing a salesman and a much greater number of projects are covered. It is considered to be far more effective than any other method, and has the advantage that it often leads to a request to attend a meeting about the project, and moreover a meeting with the appropriate people who will make the decisions.

To ensure adeqate opportunities to quote, the telephone–direct mail approach is far more effective and far less costly than employing salesmen. It is essential when this approach results in the company's being called to another meeting about the project that the team selected to attend the meeting must be headed by the marketing director or the marketing manager. At these meetings the company is still selling its services, and opportunities can be, and sometimes are, lost because a wrong line was taken at the meeting. The approach should be to show great interest, to put the company's image over in the best light possible in relation to the particular project, not to show over enthusiasm or desperation, to find out what the consultant's problems are and to offer to help in any way open. In other words, the company should behave as though it were thinking on team lines. It follows that no promises should be made which have not the slightest chance of being fulfilled. Preferably the team should consist of marketing, estimating and production representatives, and no more than three in all, with marketing leading.

Although it has been said above that consultants should be approached by telephone and letter, it is not intended that this shall apply only to architects and engineers. Clients and quantity surveyors should be approached, especially the latter as very often they recommend firms to the designers or to the clients. Indeed good and frequent conversations with quantity surveyors are very helpful in promoting a company and in obtaining information about future opportunities.

Quite apart from following up leads obtained from publications such as the press and technical journals about forthcoming projects, there will always be some projects about which the company receives unsolicited enquiries. These may come from former clients or consultants or from new ones who may be recommended or who may have been impressed by some project recently undertaken by the company, or from the general publicity put out by the company. These should be directed to the marketing branch before being passed to the estimating branch, and it is advisable for the marketing director to make personal contact with new clients or consultants as soon as possible.

Lump sum tendering in competition

Although tendering in lump sum contracts in competition is a specialized field of activity within a company, the marketing director should at least be present when the final price to be submitted is under consideration. He should know who the competitors are and the degree of keenness they are likely to show. He should be able to indicate whether the proposed price is likely to be competitive or not and whilst the settlement of the price to be submitted is the responsibility of the estimating director or the managing director, his advice should be carefully considered.

There is one aspect of tender submission that needs attention. The price should be clearly stated and not subjected to copious qualifications. A contractor is fully entitled to qualify his price if he so desires, but doing so often results in requests to withdraw the qualifications if the price is otherwise the lowest. Moreover, the insertion of qualifications indicates to the consultant that there are likely to be problems if that contractor is appointed. If there are problems, it is better that these are discussed with the consultants before the tender is finalized. By this means, it is often possible to clear up difficulties or misunderstandings and to put forward a clear unqualified price.

Design and construct contracts

This type of contract has become more frequent in the last few years, possibly because it reduces the time scale from initiation to completion. There is little evidence that it reduces cost as a general rule, but the reduction in time can be substantial.

The decision to use this type of contract may emanate from the client, but often it starts with the estate agent or property developer during the land transaction. Quantity surveyors, too, sometimes recommend this type of contract and when they do so they become the project manager, dealing directly as the client's professional adviser and the project controller so far as the contractor is concerned. Indeed, estate agents sometimes act as project controllers and appoint the contractor who is to undertake both the design and the construction, or an architect or engineer for the design work and a contractor for the construction.

There is nothing particularly difficult about these arrangements from the contractor's point of view, whether he is to design or whether an architect will be responsible for the design.

So far as marketing is concerned, if the contractor offers the whole package he will surely have advertised this fact. This may result in enquiries from time to time, particularly from former satisfied clients for whom he may have built under a traditional type of contract. Similarly, architects who may have been commissioned on a design and build basis are likely to employ a contractor with whom they have worked satisfactorily on traditional-type contracts.

Sometimes projects are advertised on a design and build basis and in these cases contractors must apply to be considered, but there is an aspect of this which needs some care. If the contract is open to competition, it is worthwhile trying to find out how many contractors will be invited to submit schemes and prices. If it is more than three or four, the chances of success decrease and an unsuccessful tender entails costs far in excess of those involved in a traditional type contract. Design costs can be substantial and these, like estimating costs, are not directly recoverable. The best type of design and build contract is the single tender or negotiated type, as indeed this is always the best type of contract from the contractor's point of view.

Prime cost plus fee contracts

Some firms specialize in this type of contract, and have been very successful. Opportunities do arise and if the type of work and the location are within a contractor's range, they should be pursued. Seldom will they be other than in competition, and until a contractor can produce evidence that he has successfully carried out this type of contract, he will find it difficult to obtain an opportunity to quote. Until he has had the experience also, he will find it difficult to be competitive, but that should not prevent a contractor quoting when the chance occurs, and he will learn from this experience.

Management contracts

These contracts are becoming more frequent. The architect carries out the design work, or most of it, and then engages a contractor to undertake the complete management of the construction work. The appointed contractor does not himself carry out any of the construction work, but is responsible for the site offices and administration as if he were the main contractor. He may or may not be responsible for some of the design work. His responsibilities are to appoint other contractors to carry out the various elements of construction and he has to co-ordinate these contracts and to produce the completed contract on schedule. For his services he receives a fee dependent on the scope of his job, but usually quoted in competition with others. Sometimes enquiries for this type of contract are unsolicited, sometimes they are advertised. Contractors who have undertaken very large contracts with many sub-contractors undoubtedly have the necessary management capability to carry out management contracts, but until a contractor can produce evidence that he has successfully undertaken a management contract, he will find it difficult to obtain opportunities to quote. Contractors interested in these types of contract must advertise that they provide this service and hope that an opportunity will present itself. They must, of course, pursue any possibilities that may arise.

Legal advice

If any contract other than the JCT or ICE types are contemplated, it is essential to understand clearly the responsibilities of the contractor. In design and construct contracts, the responsibilities of the contractor can be very onerous and legal advice is necessary before becoming a party to such contracts. The contractor may be liable not only for faulty workmanship, but also for faulty design and material failures. The liability can extend over very long periods and after obtaining legal advice, it is essential to consult the company's insurers to ascertain that adequate cover is available against these risks.

Chapter 8
The contractor as a developer

For very many years some contractors have operated as house builders on a speculative basis. They have acquired land with the requisite planning approval for house building and have erected houses for sale. Some have been extremely successful in the post-war demand for houses for owner-occupation. In 1980, 55% of houses and flats were owner-occupied, a big increase compared with 1939 and a proportion that is still increasing.

This book does not deal with house building, the marketing of which is highly specialized, but there are other types of speculative construction in which more and more authorities and contractors are becoming involved.

There is little doubt that factories and warehouses in the United Kingdom are, to a large extent, out of date and, in some cases, beyond the possibility of modernization. Many of them were built nearly 100 years ago and are sadly deficient in the layouts and the amenities necessary for modern production. The facts of the critical state of industrial buildings was highlighted in the National Economic Development Office report on Construction for Industrial Recovery. What is more, the introduction of automation and computer-controlled techniques in manufacturing and storage have proved beyond doubt that only a massive rebuilding of outmoded premises will really create production and storage facilities appropriate to modern times, not to mention working conditions acceptable in the light of safety, health and welfare requirements currently statutory. If there is to be the rejuvenation of British industry which is so frequently mentioned in political speeches by economists and, indeed, by leaders in industry, there will have to be some large-scale expenditure in creating these modern industrial buildings and the services which they require.

Indeed, much has already been done. The new towns have been developed and include areas set aside for industry where modern and attractive factories and storage premises have been built. Alas, they are not all occupied. English Industrial Estates, Welsh and Scottish Development Agencies have built new factories in strategic areas to induce industrialists to move there to help combat unemployment resulting from the decline in the heavier basic industries. Local authorities have similarly embarked on the development of industrial estates which are now to be found in most areas of the country. Private developers using their own finances or funds borrowed from such sources as the institutions and pension funds have entered into industrial estate development with enthusiasm, and some have been very successful in doing so. Others have found that the financial problems involved were beyond their capabilities and have foundered. Every day some contractor or other tackles this type of development, and in doing so puts his finances at risk. Some will be successful, others will not. It is worthwhile examining briefly some of the problems and the best courses to take to reduce the risk of failure.

Industrial estate development requires expertise not normally found in a contractor's organization. It is a marketing matter and a specialized one. It is more than assessing demand and supply; it is almost an act of faith.

The first problem in a contractor's organization is to decide whether speculative industrial development shall be undertaken by the same company which operates as a general contractor, or whether a separate company shall be formed for the purpose. There is very little to be said for the first proposal. It would weaken the contracting organization, which should be seeking to expand, and it would introduce a complication about priorities. In the end it would tend to result in a specialist division employed full-time on estate development and using staff formerly employed on general contracting. This leads to the conclusion that a separate development company should be formed with its own staff and directors and if the general contracting company's services are required for erection, they should do so in competition with the other contractors. Alternatively, the new development company should recruit its own production, estimating and design staff. These organizational problems require detailed analysis and separate costing arrangements or it will be impossible to assess whether estate development is profitable or otherwise.

If it be decided to form a separate company for industrial estate development, one of the first appointments to be made must be a marketing director, and secondly a land expert, preferably an estate surveyor member of the Royal Institute of Chartered Surveyors. If one can be found with previous experience in a development company, so much the better. Alternatively, an estate agent can be commissioned as a consultant.

In assessing demand, a number of problems arise. Demand for what: factories, storage? What services will be needed? Sewage disposal, water and power supply, effluent disposal? Where is the demand? Does it exist? Can it be created? Location? What sort of area? What sort of communication facilities? Availability of labour for employment? Skilled, semi-skilled, unskilled? Male or female?

It is of little use saying here is some land available for which planning permission for industrial development can be obtained. There is much more to it than that. Is the location one that would attract industrialists? What government or EEC grants or concessions are likely to be available? What size of premises are in demand or likely to be? What will the land cost? Will it be within the company's own financial resources or will funds have to be borrowed? If so, at what rate of interest now and how will this be likely to vary? This is no guessing game. It calls for expert opinions and judgements of the highest order. Who, in a general cotnractor's organization, can advise? Not many, if any, unless specially appointed.

Assuming that a site has been chosen after very careful consideration and acquired, a decision is then required as to how it shall be developed. First the service roads and the services must be provided and the site divided into plots for development. Next a decision is required whether to develop the whole site or part of it and whether to offer for sale or rent at least the completed shells of the buildings, leaving the internal planning to the incoming occupants. Alternatively, it may be decided not to erect any buildings until an occupant has agreed to lease or purchase and then to build what he wants. It is not advisable to build standard factories identical in size and shape. Different industries have different needs. Some firms start small and, after a time, require more space, and the plan must take account of this. Unless great care is taken in planning how best to develop the estate, it may be only partially occupied for some considerable time, with interest to pay on borrowed capital or loss of interest if one's own funds are used.

Like all development endeavours, the financial risks in industrial estate development are high. The wrong site, the wrong time, the wrong price, over-borrowing, can all have very serious effects on a company's stability. On the other hand, a successful venture can lead to profitability levels unlikely to be attained in general contracting. If this sort of venture appeals to a company, there is an opportunity to make substantial profits, but the risks are great. No company should enter this field without a great deal of thought and without ensuring that expert advice is taken.

The general principles of marketing apply to this sort of endeavour in much the same way as they do to general contracting. Promotional activities include advertising, production of brochures, sales negotiators and above all a planned programme covering finance, construction and disposal should be prepared and adhered to as far as possible. Like all programmes, some amendments will be necessary with the passage of time and this means careful monitoring of progress. This work should be in the control of the marketing director.

Chapter 9

Marketing for the small contractor

Most of what has been said earlier about the concept of marketing is relevant to the small contractor whose area of operation is very limited and whose capacity to undertake work may be restricted to projects of a few thousand pounds. The application of the principles of marketing will be very different from what goes on or should go on in medium-sized and large firms.

There is not much likelihood in a small firm of the appointment of a marketing director or even a marketing manager. The marketing function will, of necessity, fall to the managing director or to another director or partner who will have other responsibilities. Nevertheless, the small contractor must give consideration to the present position of the firm, and to its future aspirations. There must be an attempt to plan for growth, and to decide the rate of growth likely to be possible, bearing in mind finance and other resources available or which could be made available.

This necessitates a study of the market, small though it may be. In the small firm, much of the business obtained will come from personal contacts with clients who have been well satisfied with work done for them. Other work may have come as a result of tendering to the local council or to public authorities in response to advertisements or as a result of approaches from the authorities for very minor works. Is it necessary for small firms to consult planning lists or other sales leads publications to seek opportunities to tender? Is it necessary to call on architects to enquire about forthcoming projects? In some cases the answer is that either of these practices is quite unnecessary as sufficient orders come along from contracts made at the local football ground or golf club. Much depends on the desire to expand and the rate of expansion planned. Small contractors tend to increase their turnover not by taking on more projects, but by taking on larger projects and it is suggested that this is the best way forward. The projects need not be technically more difficult, although they will become so in time, but simply involve more volume of construction.

Advertising for a small contractor requires careful consideration. His best advertisement is his board displayed prominently wherever he is working. So far as advertising in the media is concerned, Yellow Pages or Thomson Local Directories are the obvious places because his clients are likely to refer to them if they have an urgent job to be done. Other advertising should be confined to local newspapers, church magazines and the like. No doubt the small contractor will have become a member of a trade organization and an active part in its activities is another way of meeting people and promoting the firm.

Public relations are as important to small firms as they are to large ones. In the case of small firms, the proprietor or managing director will most likely be known personally by his clients and their architects or engineers. If not, then every effort should be made to meet them at the earliest opportunity. Discussions about projects will frequently be between the managing director and the client or his consultant and it is important that these shall be on a friendly basis. These friendships can be fostered by frankness in discussions, strict adherence to promises made, and promptness in rectifying any mistakes which may be made. Nothing annoys a client more than work which does not start on the date agreed, and which takes much longer to complete than was originally promised. It is important to build up confidence in clients and consultants. Consequently promises which cannot be fulfilled should never be given.

The indications of efficiency should be kept up by the prompt rendering of accounts clearly stating what is owed and why. Accounts which cannot be easily understood by the recipient lead to requests for clarification and can lead to loss of clients. It is important, therefore, to attach to the copy account details of costs involved so that queries can be dealt with promptly. It always pays the managing director or one of his senior staff to visit the client on completion of the work to ensure that the client is satisfied. The work is important to the client and a visit from the firm will demonstrate that it was important to the firm also. These visits could result in additional orders or opportunities.

100

Opportunities for growth

There is little doubt that householders are spending more time in their homes than they did before the availability of television and this trend seems likely to continue as the video market develops. Consequently the occupants are for ever trying to improve the quality of their homes by the installation of double glazing of their windows, the provision of more power points to match the increasing use of electrically powered accessories to easier living. In kitchens there have been big improvements made by the installation of fitted kitchen units and this is still increasing in frequency. It necessitates new or revised plumbing arrangements to serve washing machines, different sink and dishwashing facilities, and electric wiring changes too. It usually involves wall-tiling and inevitably redecoration.

In bathrooms too, improvements are being made in the facilities with such installations as showers and bidets and the replacement of baths, wash basins and WCs with more modern versions.

Extensions are in demand to form utility rooms or to enlarge sitting rooms or dining rooms, or to provide single or double garages.

All these demands provide work for the building industry and particularly for general contractors who employ all the requisite tradesmen, although there are some firms who specialize in the various trades who undertake part of the work in conjunction with other specialists who do the rest.

There is no reason for small general contractors not to seek work of this kind. There is plenty of it available from private householders. What is required is a fair price, promptness in starting and finishing, and good quality work. Advertising that the service is available will soon result in enquiries.

Local authorities have large contracts available for these kinds of work to be carried out and contracts can cover up to one hundred of their houses. This is repetitive work which requires skilful organization and can lead to useful and profitable expansion of a company.

There is also refurbishing of existing buildings either to improve general conditions or often to enable a change of use. Buildings formerly in use as cinemas or music halls have been converted to other uses such as factories, warehouses or offices and a good deal of work has been necessary to achieve these results. This sort of work is frequently available, as also is work in existing factories where production layouts need to be changed to match more modern machinery.

The above examples of work availability resulting from householders seeking more comfort or the increasing use of labour saving devices, and the modernization of other buildings or their change of use are by no means comprehensive. They are merely indications of areas of activity which small contractors would be well advised to consider in their marketing management discussions.

Chapter 10

Contracting overseas

Although by far the majority of firms in the British construction industry will operate only in the United Kingdom, and many of them in localities not far removed from their head offices, there will be a growing number who will be interested in markets outside the United Kingdom. There are attractive markets in many overseas countries where some British companies have already operated with varying degrees of success, and there are others which have not yet been fully explored.

Companies planning for future profitable growth should be considering the possibility and desirability of entering the large overseas markets available, many of which seem likely to expand in the future. Four areas are attractive: North America, Europe, the Middle East and the Far East. In these areas demand for constructional services is high and such restrictions as there are can be accepted by British firms without serious concern. Progress in these areas for British firms will involve capital expenditure in some way: purchase of assets, partnership with indigenous companies, establishment of offices and depots, or the formation of specialist subsidiaries.

It is important to appreciate that overseas work at best can provide only a fraction of a firm's total turnover, but a fraction with the distinct possibility of growth. It follows that successful overseas operations are unlikely unless there is a sound United Kingdom base.

Before any decision to attempt to enter an overseas market is undertaken, a thorough examination of the structure of the area, its politics, its stability or otherwise, its forward programme of development, its practices and procedures so far as construction is concerned needs to be undertaken. A foreign country's history, culture, geography, climate, religion, laws and living standards are all very different from anything experienced in the United Kingdom. Furthermore, although modern society in the United Kingdom has been developed over the last 100 to 200 years, some overseas countries, particularly in the Middle East, are trying to achieve such changes over a much shorter period. They are eager for quick results and the oil-rich countries have the finances to back up their policies.

So far the bulk of the effort by British contractors and consultants has been directed to the Middle East: Saudi Arabia, the United Arab Emirates, Kuwait, Bahrain, Iraq and Egypt. About 40 to 50% of overseas work carried out by British contractors since 1973 was in these areas. There are other countries where demand exists, notably the Sudan, Oman, Jordan, Algeria, Morocco, Syria and Tunisia, and very sizeable demand too. With the reduction in United Kingdom demand and the difficulties of obtaining satisfactory profit margins these markets have attracted British contractors, particularly the large ones, and the overseas countries have not been slow to encourage them to undertake work there.

Those companies that embarked on overseas commitments in the early 1970s considered they had struck gold. The procedures were such that there were opportunities to obtain profit margins far in excess of those attainable in the United Kingdom. As time went on the overseas countries learned a great deal about contracting procedures, and there was a tightening up and an approach which made some overseas procedures much more onerous than in the case of United Kingdom contracts. Added to this was the increased competition from American, Italian, Dutch, Indian, Pakistani, French, German and Korean firms. The result was a reduction in margins and the market became considerably tougher. There are opportunities in the Far East where competition is also keen and in the Western states of North America. There are vast undertakings being planned in countries bordering the Mediterranean, and, of course, in the African continent further south. These are well worth consideration in detail.

Before deciding to try to obtain overseas contracts, it is vital that the marketing director and production director should visit the country chosen. They should stay there for as long as is necessary to discover precisely what life is like there and how the company can fit in with the constructional activities and procedures. They should consult the British Overseas Trade Board in London for a list of people they should meet, and they must make contact with the British government's trade commissioner and other government officials. There are likely to be offices of British consultants and other British contractors in the country, and valuable advice and help will be available from these sources. Visits must also be made to local contractors as it will almost inevitably be necessary to enter into partnershup with one of them before contracts can be sought.

There are many questions that will need answers before decisions can be taken, and these include:

Is there anyone in the company with overseas experience?

How much is known about the problems of overseas trade? Who in the company will deal with export credit, currency exchange, shipping and freight costs, insurance, air travel and accommodation?

What kind of staff will be needed at home and overseas to prepare estimates and to manage contracts?

How many staff and operatives will be prepared to undertake an overseas tour, and on what terms?

What about wives and children, leave, accommodation, education for children?

Is sufficient known about materials prices and availability and labour rates to enable sound estimates to be prepared?

How will staff repatriated at the end of a tour be employed? (They will have grown in stature and experience and will expect higher status than their previous UK posts.)

Will communications with overseas staff be satisfactory?

There is an organization in London known as Building Marketing Consultancy Ltd which has produced in-depth studies of overseas markets and these reports are available for a fee. It is strongly recommended that they are studied in detail before any decision is made to try to enter overseas markets.

Generally in Middle East countries British contractors can only operate as junior partners in a locally-based company, or as sub-contractors. It follows that great care must be taken before entering into such a partnership and it is advisable to talk to several firms before commencing serious negotiations. It may be found that the local firm's main contribution to the joint effort will be in the provision of information about future opportunities for contracts, and about likely tender levels. The UK firm's major contribution will be to provide the original estimate and, if successful, to undertake the management of the contract with particular reference to quality of work and speed of erection. This management function will include recruitment and welfare of the labour force, ordering and progressing of materials, appointment and programming of sub-contractors.

If a partnership is entered into, it will be necessary to send out administrative staff as well as production staff to manage contracts. Plant and transport may be sent out also, but this can be obtained locally. It will be necessary to find accommodation for staff and possibly their wives and children. Schools exist which are reported as satisfctory, although in some cases children are placed in schools in the UK at the company's expense. These and many other costs have to be written into tender prices so they appear to be very large in comparison with the cost of similar projects in the UK.

The principles of marketing apply to overseas work as they do in UK work, but there are problems not encountered in the UK. Certainly annual and long-term target turnover should be planned and types of work to be aimed for chosen. Since the UK company is only a partner, agreement with its partner will be necessary, and this may not always be easy to achieve.

Company promotion through advertising, brochures and sales literature should follow UK lines. There are many advertising outlets through the press and technical journals and it should be remembered that UK-based consultants will be operating in overseas countries. Consequently advertising in the UK about a company's operations overseas is not out of place. Free publicity can be obtained through the Central Office of Information who are always prepared to include articles about UK contractors' activities overseas in their publications.

Regular reports from staff located overseas should be sent to the marketing director who should extract those items worthy of publication and see that press releases are issued.

Brochures of overseas work should be built up as contracts are awarded and, of course, they should be printed in the local language as well as in English. They should be distributed to government departments in the country concerned, to the British embassy or legation, and to consultants involved in work in the country.

Prequalification documents for overseas contracts are far more detailed in some respects than for UK contracts. Much detailed information about qualifications and experience of staff is required and this must be prepared and copies kept for use in future applications.

British Overseas Trade Board

The British Overseas Trade Board was established in 1972 to help British exporters. Its members are drawn from government and industry under the presidency of the Secretary of State for Trade, and the chairmanship of a prominent industrialist.

The Board is always ready to provide information, advice and assistance to exporters, and those in the construction industry contemplating operations overseas would be well advised to consult the Board.

The Board operates an export marketing research scheme which provides marketing information for contractors and others for many areas of the world. The intention is to persuade exporters to undertake their own marketing research, and in approved cases the Board will make grants up to one-third the cost of a contractor employing a professional consultant for this purpose. Where the research is undertaken 'in house' a grant of 50% of travelling costs, plus a daily allowance for time spent overseas, is available in approved cases. There are also grants of up to one-third the cost of setting up a market research department, or for the purchase of published market research reports.

The Market Advisory Service operated by the Board provides information about the likely demand for construction services in overseas markets, and advice on how to obtain suitable local representation. This advice is very useful to contractors considering the market potential of overseas areas before undertaking visits to the areas themselves.

The Export Intelligence Service of the Board provides regular market information, including calls for tenders to subscribers, and a fee is charged for this advice. This service is certainly worth having, especially in the early days of a firm's overseas involvement. Once the firm is established overseas its own members on the spot will be able to obtain marketing information. The handbook BOTB Services available from any of the regional offices of the Department of Trade should certainly be studied by contractors contemplating operations overseas. It sets out clearly the many ways in which the Board can assist.

Export Group for the Constructional Industries

Consideration should be given to becoming a member of the Export Group for the Constructional Industries. This group was formed in 1940 to promote interest in design, construction and supply of all types of constructional projects outside the United Kingdom. It is a non-profit making trade association financed by its members.

Services to members include a weekly *Commercial Bulletin* with details of projects overseas, periodically a *Market Information Bulletin*, and an *EEC Bulletin* which contains information about EEC activities likely to affect the activities of members. The group also publishes a well-produced and illustrated brochure of the services offered by members and this is widely circulated amongst likely clients and their consultants.

Contractors who are interested should obtain a copy of *The Export Group for the Constructional Industries: Aims and Activities*, available from the Group's offices at 2 Dean Trench Street, Smith Square, London SW1 3HD. They will also find the Group's Annual Report interesting and informative.

The European Economic Community

The United Kingdom's entry into the Common Market has made contracting in member countries somewhat easier. A directive of the EEC made it necessary for all publicly financed projects in excess of a certain value (at present £660 000) to be advertised in the *EEC Journal*. Any firm in any of the EEC countries is free to make application for inclusion in the tender list for any project so advertised. Strangely enough, not many European firms have secured contracts in the United Kingdom as a result. It is not known how many have sought work here. Equally, not many British firms have sought or at least obtained contracts on the continent of Europe. Those firms that are interested in working on the continent would be well advised to consider forming a partnership with a local firm in the country of their choice. The procedures are different, the industrial agreements are different, and local knowledge is invaluable.

It is, perhaps, not as widely known as it should be that EEC development grants are available for construction projects not only in EEC countries, but in more than 60 overseas developing countries in Africa, the Caribbean and the Pacific. There are funds available for both public and private developments in the United Kingdom and many projects have been undertaken here funded wholly or in part by the EEC. Similar arrangements are available in countries outside the EEC and any firm interested should get in touch with the First Secretary (Commercial) in the office of the UK Permanent Representative to the European Communities in Brussels, or with the Department of Trade, Victoria Street, London, SW1. They should obtain copies of *Business Guide to the European Development Fund*, available from the British Overseas Trade Board in London, and the guide *How to participate in contracts financed by the European Development Fund*, published by the EEC and available from the Regional Export Offices of the Department of Industry and Trade. Information on individual projects is published in the 'S' series supplement to the *Official Journal of the European Communities* obtainable from Her Majesty's Stationery Office.

Export Credits Guarantee Department

Contractors contemplating involvement in overseas contracts will be well advised to consult the Export Credits Guarantee Department. This is a governmental department which helps exporters of goods and services by insuring them against the risks of not being paid. The risks covered in this way are set out in the publication ECGD Services, and other publications which are available from the Department's Head Office or any of its regional offices.

A premium is charged for this service and is calculated broadly according to the nature of the risk (including the market) and the length of the credit period.

The Department also gives guarantees to support finance provided by UK-based banks and the issue of performance bonds where these are required by the client.

The full details of services available from ECGD are set out in the publication referred to above, and contractors are advised not only to study it, but to discuss their particular proposals for exporting with the officials of the Department, before they enter into commitments.

Costing

Earlier in this book there has been a reference to the need for a proper staff-costing procedure. This is essential if proper analysis of costs is to be undertaken. It is even more important so far as overseas work is concerned where costs are inevitably much higher than in the case of work in the UK. If possible, a complete overseas department should be set up so that all the staff therein are employed full time on overseas work. In this way the staff become part of the overseas company or companies, and there can be no doubt as to where their costs are to be allocated.

If this cannot be done because the amount of overseas work does not warrant it, then staff will find themselves working part of the time on UK problems and part on overseas work. Their costs must be split as accurately as possible between UK and overseas, otherwise it will never be known what the relative profitabilities are. Overseas activities should, preferably, be organized in a way similar to a UK region so that a clear picture of success or otherwise is available.

The world has become much smaller with the development of high speed air travel. Contractors can start by being local firms, grow to be national firms operating anywhere in the United Kingdom, and many have developed in this way. The next logical development is to international status operating in some well-chosen and profitable overseas markets. Progress in this direction is far from easy and requires sound judgment if it is to be successful. This chapter has attempted to draw attention to markets available and to some of the organizations that can be of assistance to contractors contemplating expansion in this direction. The chapter is not comprehensive, but it is intended to indicate some of the ways open to those who might wish to achieve international status. The competition is keen: the experience and the profits may be rewarding.

Bibliography

Books

Calvert, R. E., *Introduction to Building Management*, 4th edn, Butterworths (1981)
Elvy, B. Howard, *Marketing made Simple*, 2nd edn, W. H. Allen (1980)
Jepson, W. B. and M. P. Nicholson, *Marketing and Building Management*, Medical & Technical Publishing Co. Ltd.
Smallbone, Douglas, W., *An Introduction to Marketing*, Staples Press (1968)
Wilson, Mike, *The Management of Marketing*, Gower (1980)

Papers and reports

Banwell Committee, *The Placing and Management of Contracts for Building, Civil Engineering Works*, HMSO (1964)
Bell, R., *Marketing and the Larger Construction Firm*, Occasional Paper No. 22, Chartered Institute of Building (1981)
British Overseas Trade Board, *B.O.T.B's Services*
Corser, T. D., *The Effect of Membership of the European Economic Community on the British Construction Industry* Department of Transportation and Environmental Planning, University of Birmingham
Department of the Environment, *Housing and Construction Statistics 1971–1981*, HMSO
Economist Intelligence Unit Ltd., *Public Ownership in the Construction Industries*
Export Credits Guarantee Department, *E.C.G.D. Services*
The Export Group for the Constructional Industries (2 Dean Trench Street, Smith Square, London SW1 3HD), *Aims and Activities*
The Export Group for the Constructional Industries, *41st Annual Report 1980–81*
Working Party, NFBTE, *Marketing for the Building Industry* (July 1970)
Working Party Report on Building, HMSO (1950)

Regular publications

Building (weekly)
Contract Journal (weekly)
Department of the Environment, *Housing and Construction Statistics Quarterly*, HMSO
Department of the Environment, press notices
Joint Forecasting Committee of the Building and Civil Engineering EDCs, *Construction Forecasts* (2 per year), HMSO

Index

Acquisition of companies, 43
Advertising, 44–61
   agencies, 60–61
   annual, 47–48
   budgets, 56
   display, 52–54
   effectiveness of, 51–52
   hoardings, 58–60
   radio, 57–58
   series, 48–49
   support, 54–56
   television, 58
Approved lists of contractors, 13
Areas of operation, 27

Banwell Report, 14
BOTB, 106
Brochures, 79–81

Cash flow, 17
Communications, 62–64
Company image, 45
Company reports, 72–77
Construction for industrial recovery, 94
Consumer goods, industries, 12–13
Contract awards, 68–70
   completion, 73–75
   handover, 73–75
   progress, 71–72
Credit control, 19

Definition of marketing, 1
Design and construct contracts, 91–92
Development Agencies, 95
Direct mail, 90

ECGD, 108–109
Ecological pressures, 24
Estate surveyors, 96
Estimating, 40
European Economic Community, 107–108
Exhibitions, 82–83
Export Group for Construction Industries, 107

Feed back, 35–36
Films, 81–82

Growth, 17–18, 27, 35, 40, 100

Hotels, 23
House journals, 83–84

Industrial Estates, 94–97
   factors, 21–24
   Training Act 1964, 11
Inflation, 25
International scene, 24–25

Job records, 36–38

Labour only sub-contractors, 28–29
Legal advice, 93
Legislation, 21
Leisure, 23
Letter headings, 64, 85
Logotypes, 84–86
Lump sum tendering, 91

Management contracts, 93
Market research, 12
Market organizations, 33
Marketing planning, 26–30
Motorcar industry, 2

National Economic Development Office, 33, 94
National Federation of Building Trade Employers Working Party Report, 5
Number of firms, 15

Official openings, 73–75
Open tendering, 13
Organization, 29–30, 32–33
Outstations, 40–42
Overseas markets, 103–106

Photographs, 50
Placing of contracts, 13
Planning, 8–10
Planning factors, 36
Plant, 28
Political decisions, 21
Presentations, 77–78
Press conferences, 77
Press notices, 65–74

Prime cost contracts, 92
Profitability, 4, 17, 18
Public relations, 62–85
Public relations consultants, 64–65

Regional organizations, 40–43

Sales leads, 33
Sales literature, 79–81
Salesmen, 86–89
Satisfaction of employees, 20, 46, 62
Schools, 23
Shareholders, 19, 62
Size of firms, 16
Shopping, 23
Social factors, 21–24
Sponsorship, 78–79
Sporting events, 78
Staff associations, 20
Structure of construction industry, 14–16
Supply and demand, 6–7
Suppliers, 19

Telephone procedure, 63–64
Telephone sales, 89–90
Tendering, 35
Thomson directories, 99
Topping out, 73
Turnover, 17–18, 27–29, 34–36
Types of work, 27–28

Visits to consultants and clients, 90

Working Party on Building Report, 5

Yellow Pages, 99